Circadian Rhythms for Future Resilient Electronic Systems

Xinfei Guo • Mircea R. Stan

Circadian Rhythms for Future Resilient Electronic Systems

Accelerated Active Self-Healing for Integrated Circuits

Springer

Xinfei Guo
University of Virginia
Charlottesville, VA, USA

Mircea R. Stan
University of Virginia
Charlottesville, VA, USA

ISBN 978-3-030-20053-4   ISBN 978-3-030-20051-0   (eBook)
https://doi.org/10.1007/978-3-030-20051-0

This Springer imprint is published by the registered company Springer Nature Switzerland AG.
The registered company address is: Gewerbestrasse 11, 6330 Cham, Switzerland

To our families and friends...

Preface

The downscaling of CMOS technologies in the semiconductor industry has continuously offered better performance, lower power, and higher levels of integration. However, in advanced nodes with smaller feature sizes, on-chip components such as transistors and interconnects are experiencing more aggressive degradation due to wearout (aging) effects, which are dominated by bias temperature instability (BTI) and electromigration (EM). Transistors become more susceptible to voltage stress due to the increased effective field with the scaling of the thin oxide. Similarly, the shrinking geometries of metal interconnects result in higher current densities, while the tremendous number of transistors within a compact area results in higher power densities as well. Together, these lead to increased on-chip temperatures which further accelerate the wearout effects. In the meanwhile, with the ubiquity of electronics (e.g., Internet of Things and edge devices) in our daily lives, there have been increasing demands for reliable system design. Many of such applications require very long lifetime, higher utilization rate, and tighter hard error tolerances. The systems are possibly deployed in extreme environmental conditions, such as high temperatures, which, unfortunately, further accelerate wearout.

Conventional techniques of coping with wearout by "tolerating," "slowing down," or "compensating" still leave the wearout issues themselves unresolved since they keep accumulating as the system operates. This book introduces a new category of techniques that can "repair" wearout in a physical sense through accelerated and active recovery, allowing wearout (both BTI and EM) to be reversed by actively applying several techniques, such as negative voltages (for BTI) and reverse currents (for EM), with temperature possibly acting as an accelerating factor, thus leading to effective *accelerated active self-healing*. In this book, we aim to present our key discoveries on recovery behaviors for several dominating wearout mechanisms such as BTI and EM. By experimentally studying the frequency dependent behavior of wearout and recovery, we demonstrate that even the permanent components of wearout can be almost fully eliminated and avoided by using in-time scheduled recovery. The topics cover recovery theory, experimental results, circuit-architecture system (cross-layer) design techniques that enable accelerated self-healing on chip, and wearout in emerging applications, such as in IoT edge devices. In addition, we present our findings from a comprehensive study of a leading-edge technology node (e.g., FinFET) vs. other technologies.

The book is divided into five parts. Part I, which contains Chap. 1, gives an overview of CMOS and interconnect wearout and conventional mitigation techniques. As the main focus of this book, the concept of accelerated active self-healing is also introduced in this part. Part II, which contains Chap. 2 that presents transistor wearout BTI accelerated self-healing results and Chap. 3 that mainly focuses on EM recovery measurement results, presents detailed theories and experimental results of recovery behaviors. In both chapters, the analyses of potential benefits enabled by accelerated active self-healing are detailed as well. To instrument the concept of on-chip recovery and to fully utilize the explored recovery behaviors that are presented in Part II, we propose several implementation solutions across the system hierarchy, and these are discussed in Part III, which includes Chaps. 4 and 5. In Chap. 4, a full set of circuit IP blocks, including recovery boost components, novel BTI and EM sensors, multimode recovery assist scheme, and novel power-gating structures, are designed and implemented. As wearout-induced failures become more visible at the system level, we also explore several potential architecture and system level solutions that are able to take advantages of intrinsic sleep behaviors for full recovery, and these are detailed in Chap. 5. Overall, the techniques presented in this part can work together to guarantee that the entire system performs for more of the time at higher levels of performance and power efficiency by fully exploiting the extra opportunities enabled by the accelerated active self-healing. As leading-edge nodes such as FinFETs endeavor to offer advantages of future scaled devices while offsetting the problems introduced by many generations of planar CMOS scaling, adapting to the new challenges and fully benefiting from FinFETs require new knowledge and design experiences. To contribute to this knowledge base, in Part V (Chap. 5) of this book, we firstly present a comprehensive study based on circuit simulations across multiple technology nodes ranging from conventional bulk to advanced planar technology nodes such as fully depleted silicon-on-insulator (FDSOI), to FinFETs. As challenges such as wearout appear to be more pronounced in these advanced nodes, this can grow to be critical especially in IoT applications in which the industry is in the process of introducing new technologies. Each of these end markets has unique needs and characteristics, which affects how chips are used and under what conditions. We investigate how wearout can impact different categories of future IoT applications with foundry-provided wearout models as the second main component of this chapter. We conclude that wearout needs to be considered in the full design cycle and the IoT lifetime estimation requires to incorporate wearout as an important factor. Potential IoT-specific design solutions for mitigating wearout are also presented in this part.

The intended audiences of this book consist of researchers and designers in the yield, circuit and system reliability, digital design, and EDA tool development areas. The book is also of interest to graduate students or senior undergraduate students who are interested in learning advanced technology nodes and are willing to embrace the design-for-reliability challenges. The contents covered in the book reflect our 10-year research efforts in this field. We wish that readers across disciplines find the material in this book relevant and gain an understanding and appreciation of this exciting field. We will certainly appreciate any feedback or comments.

Charlottesville, VA, USA   Xinfei Guo
Charlottesville, VA, USA   Mircea R. Stan
March 2019

Acknowledgments

Many people have helped make this book possible.

First of all, we are very grateful to Dr. Xinfei Guo's Ph.D. dissertation committee members from the University of Virginia and IBM Research, as the major material of this book is based on his dissertation. Specifically, we want to thank Prof. Kevin Skadron, Prof. James Aylor, Prof. Samira Khan, and Prof. John Lach for their technical suggestions and comments and for their challenging questions. We would like to thank Dr. Pradip Bose from IBM Research for his kindness, support, and helpful comments through the whole project.

We appreciate all the feedback and comments from our collaborators and industry liaisons, namely, Prof. Wayne Burleson (UMass, Amherst), Dr. Tanay Karnik (Intel), Dr. Sudhanva Gurumurthi (AMD & UVa), Dr. Matthew Ziegler (IBM), and Prof. Mohamed El-Hadedy (Cal Poly), for this research. We would like to express our gratitude to colleagues and alumni from the High-Performance Low-Power (HPLP) lab at the University of Virginia who laid solid foundations for this research and helped at different stages of the project. Dr. Alec Roelke helped with the cross-layer implementations; his simulation framework "OldSpot" has made it possible for us to explore the opportunities enabled by accelerated active self-healing at the architecture level. Vaibhav Verma, Patricia Gonzalez-Guerrero, and Sergiu Mosanu helped with simulations for Chap. 6, Dr. Kaushik Mazumdar provided insights on circuit-level solutions, and Dr. Linqiang Luo helped with the wire bonding for the EM recovery experiments. The contributions of these smart researchers and collaborators led to the development of this book.

Our research ideas discussed in this book couldn't have been developed without the inspiration from the research products of other pioneers who have left their marks in the field of CMOS and interconnect wearout, including Prof. Muhammad Ashraful Alam (Purdue), Dr. Shekhar Borkar (Qualcomm) , Dr. Pradip Bose (IBM Research), Prof. Yu Cao (ASU), Prof. Jörg Henkel (KIT), Dr. Vincent Huard (STMicroelectronics), Prof. Souvik Mahapatra (IIT), Prof. Mehdi Tahoori (KIT), Prof. Sheldon Tan (UC Riverside), and many many others. We take this opportunity to thank them for all of their great research work that helped us understand the fundamentals and inspired our research ideas.

We would also like to thank all the funding sources that supported our research over the years: NSF (CCF-1255907 and 1543837); SRC (GRC program Task 2410.001); the Center for Future Architecture Research (C-FAR), one of the six SRC STARnet centers, sponsored by MARCO and DARPA; the Center for Research on Intelligent Storage and Processing-in-memory (CRISP), one of the six SRC JUMP centers; and the IEEE Circuits and Systems Society (CASS). We are grateful to all the reviewers for our papers for their questioning and feedback. We also thank Mr. Charles Glaser from Springer and his team for their strong support and quick actions that made a speedy and timely publication possible.

We want to thank our families and friends who offer endless love and support to us all the time. We would like to dedicate this book to them.

Last but certainly not least, we want to thank you, our readers. We sincerely hope that you will find perusing this book as exciting and informative as we have intended.

Charlottesville, VA, USA   Xinfei Guo
Charlottesville, VA, USA   Mircea R. Stan
March 2019

Contents

Acronyms

AAR	Accelerated active recovery
ABB	Adaptive body biasing
ADE	Analog design environment
AR	Accelerated recovery
AS	Accelerated stress
ASIC	Application specific integrated circuit
BEOL	Back end of line
BOL	Beginning of lifetime
BRAM	Block random-access memory
BTI	Bias temperature instability
CLASH	Cross-layer accelerated self-healing
CLM	Channel length modulation
CMOS	Complementary metal-Oxide Semiconductor
CUT	Circuit under test
Decap	Decoupling capacitor
DFT	Design for test
DIBL	Drain-induced barrier lowering
DRC	Design rule check
DVFS	Dynamic voltage and frequency scaling
EDA	Electronic design automation
EDP	Energy-delay product
EM	Electromigration
EOL	End of lifetime
EOT	Equivalent oxide thickness
EPROM	Erasable programmable read-Only memory
FBB	Forward body bias
FD-SOI	Fully depleted silicon on insulator
FEOL	Front end of line
FinFET	Fin field-effect transistor
FO4	Fanout of 4
FPGA	Field-programmable gate array

GIDL	Gate-induced drain leakage
HCI	Hot-carrier injection
HP	High performance
HPC	High-performance computing
I/O	Inputs/outputs
IC	Integrated circuit
IG	Independent gate
IMP	Average performance improvement
IoE	Internet of Everything
IoT	Internet of Things
IP	Intellectual properties
IR	Irreversible wearout
IR-Drop	Electrical potential difference between the two ends of a conducting phase during a current flow. This voltage drop across any resistance is the product of current (I) passing through resistance and resistance value (R)
ISA	Instruction set architecture
LELE DP	Litho-etch-litho-etch double patterning
LER	Line-edge roughness
LUT	Look up table
LV	Low voltage
MC	Monte Carlo
MCU	Microcontroller unit
MEOL	Middle-end of Line
MRAM	Magnetic random access memory
MTJ	Magnetic tunnel junction
MTTF	Mean time to failure
MUX	Multiplexer
NBTI	Negative-bias temperature instability
NMOS	N-channel metal-oxide-semiconductor field-effect transistor
NVM	Nonvolatile memories
PBTI	Positive-bias temperature instability
PC	Personal computer
PCRAM	Phase-change random access memory
PDC	Pulsed DC
PDK	Process design kit
PDN	Power delivery network
PLB	Programmable logic block
PMOS	P-channel metal-oxide-semiconductor field-effect transistor
PMU	Power management unit
PnR	Place and route
POI	Path of interest
PPA	Power performance area
PUN	Pull-up network
PV	Process variations

PVT	Process, voltage, and temperature variations
RD	Reaction diffusion
RDF	Random dopant fluctuations
RO	Ring oscillator
rob	Reorder buffer
RRAM	Resistive random access memory
RTL	Register-transfer level
RTN	Random telegraph noise
SADP	Self-aligned double patterning
SC	Switched capacitor
SEM	Scanning electron microscope
SG	Shorted gate
SID	Spacer is dielectric
SIM	Spacer is metal
SoC	System on chip
SOI	Silicon on insulator
SPEC	Specifications
SPI	Serial peripheral interface
SPICE	Simulation program with integrated circuit emphasis
SRAM	Static random access memory
ST	Sleep transistor
STI	Shallow trench isolation
STT	Spin-transfer torque
TCAD	Technology computer-aided design
TD	Trapping-detrapping
TEI	Thermal effect inversion
TSV	Through-silicon vias
TT	Typical-typical
TTF	Time to failure
ULP	Ultralow power
UV	Ultraviolet
VLSI	Very-large-scale integration
VS	Voltage stacking
VTC	Voltage transfer curve

Part I
Overview

Chapter 1
Introduction to Wearout

1.1 Wearout in CMOS Circuits

The never ending demands to deliver higher performance, better energy efficiency, and more integration on a single die have depended on the continuous downscaling of CMOS transistor feature sizes. Although technology scaling has been advantageous for many metrics, these advancements have also augmented the impact of reliability issues [1]. Reliability refers to the probability that a system is able to perform its intended functions for a given lifetime under given conditions. Sources of unreliability at the hardware level include variations caused by manufacturing and operating conditions, soft errors caused by electrical noise or external radiations, and wearout (aging)[1] failures that are caused by device degradation [2]. Process variations have been well studied for decades and are usually modeled accurately as part of the process design kit (PDK), which guides circuit designers to design taking them into account. Soft errors, also called transient faults or single-event upsets (SEU), may cause computation errors and corrupted data, but they are temporary and do not affect the lifetime of the computing systems [3]. Unlike the first two unreliability sources, wearout effects, which raised to prominence more recently, are manifest during the system lifetime and highly depend on unpredictable operating conditions. Wearout has thus grown to be a huge reliability threat to the lifetime of digital circuits and systems [4–6]. The reasons behind this have been manifold [7], but they can be summarized in the following two categories.

The first one is from the *technology* aspect—Fig. 1.1 shows the projected wearout acceleration across multiple technology nodes [8], and it has also been demonstrated and confirmed by recent measurement results in [9]. As technology scaling is

---

[1]In this book, the terms "aging" and "wearout" are used interchangeably. Circuit aging/wearout refers to wearout effects at both transistor level and interconnect level. Transistor aging/wearout mainly refers to BTI effect or other effects that affect CMOS transistors.

© Springer Nature Switzerland AG 2020
X. Guo, Mircea R. Stan, *Circadian Rhythms for Future Resilient Electronic Systems*, https://doi.org/10.1007/978-3-030-20051-0_1

Fig. 1.1 Projected wearout vs. technology scaling from [8]. Y-axis refers to wearout related metrics, and all values are normalized to the 32 nm node

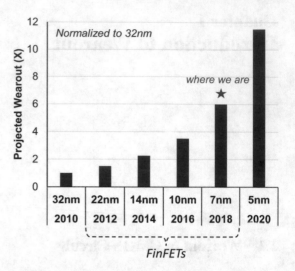

reaching the nanoscale regime [6, 10], the transistors become more susceptible to voltage stress [9–12] due to the increased effective field resulted from the reduced equivalent oxide thickness (EOT) [5]. Similarly, the shrinking geometries of metal layers result in higher current densities, and the tremendous number of transistors within a compact area leads to higher power densities as well. Together, these lead to increased on-chip temperatures which potentially accelerate the wearout effects [13]. Moreover, advanced technologies such as FinFETs have given rise to several new wearout concerns due to new effects such as self-heating [9, 14]. As technology scaling enables tens of billions of transistors to fit on one chip, the challenge is that the failure rate of single transistors needs to actually decrease so that the historical values of mean time to failure (MTTF) of the whole system can be maintained.

Besides the technology scaling factor, wearout issues also appear to be more pronounced from the application perspective [15]. In high-performance computing applications such as servers, the system utilization has been increasing significantly especially due to the advent of the cloud computing. The goal for cloud operations is to maximize utilization by balancing compute jobs across an entire data center. This means that a system may run most of the lifetime (>5 years) without stopping. The approach is energy-efficient, but it can result in very high accumulated wearout and eventually to a reduction in performance [16]. Utilization trends are shifting inside the emerging embedded applications such as the Internet of Things (IoT) edge devices and automobiles, which will continue until fully autonomous vehicles replace human drivers [17]. In wearables or medical devices, where circuits may work in near/sub-threshold for ultra-low power (ULP) operation, the sensitivity of transistor ON current to threshold voltages is much higher than in super-threshold regimes [18]. Also, demanded by marketing and applications, these edge devices usually have very strict resiliency requirements [6, 19] and require longer lifetime.

For example, some biomedical applications require a lifetime of more than 50 years for medical implants [11, 19]. Finally, wearout issues are not just about duration, they are also highly temperature-dependent. Many such systems need to operate in extreme environmental conditions, such as under high temperatures (without cooling), which, unfortunately, further speed up the wearout process [20].

Wearout is a time-dependent reliability mechanism that is caused by several physical mechanisms that conspire to worsen metrics across the system hierarchy [12, 21], with performance degradation or intrinsic faults at the circuit level [16, 22], errors at the architecture level [23], and failures at the system level [23, 24]. As shown in Figs. 1.2 and 1.3, in general, at the transistor level, also known as front-end-of-line (FEOL), bias temperature instability (BTI) is one of the most prominent wearout mechanisms [10, 12, 23, 25]. It is characterized by the increase of the absolute value of threshold voltage ($|V_{th}|$) and the reduction of the carrier mobility (μ). In metal layers, known as back-end-of-line (BEOL), electromigration (EM) is the dominant reliability threat that increases the wire resistance R over time (soft wearout), and ultimately can lead to breaks or shorts in the wire (hard failure). EM is especially critical for power delivery networks (PDN) in modern ICs [4, 13, 26, 27]. While both wearout effects happen due to stress caused by voltage or current, when the stress is removed, there is a level of recovery, but usually at much lower rate than the wearout process. In the next section, we will review some of the state-of-the-art techniques for mitigating both wearout effects.

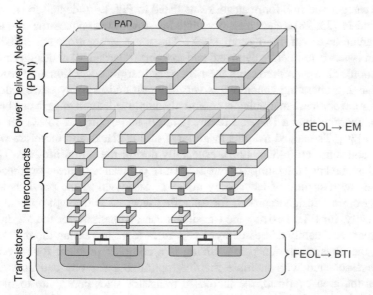

Fig. 1.2 Illustration of BTI and EM on a small section of the typical CMOS process (simplified for illustration purpose, not tied to any specific technology node). BTI occurs in transistors, and EM appears in metal wires (interconnect and power delivery network)

Fig. 1.3 Illustration of BTI
and EM in a CMOS circuit.
BTI shifts the threshold
voltage of a transistor, and
EM increases the resistance
of a metal wire

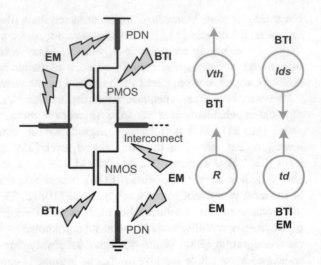

1.2   Existing Wearout Mitigation Techniques

In the past decade, various techniques have been proposed to deal with both BTI
and EM wearout issues from system level down to circuit level, and from design
time (static solutions) to run time (dynamic solutions). Overall, these techniques
can be categorized into four categories as listed in Fig. 1.4 (adapted from a survey
presented in [23, 28]), each category includes several techniques that are applied
at different levels of the system stack. The most straightforward solution for
wearout issues is to tolerate them and add margins (guardband) at design time (pre-
fabrication). However, predicting the necessary margin under dynamic workloads
and changing operating conditions is very difficult and in many cases unfeasible.
Therefore, worst-case margining is commonly employed. Such margin can be >20%
for time margins over a 10-year lifetime [21, 29], and as high as 14.5% for voltage
margins [30]. The added margins mean large timing slacks and therefore wasteful
energy consumption ($\sim$30% [28]), especially during the initial lifetime. This has
been illustrated as "tolerating" in Fig. 1.5. The significance of these overheads also
increases with scaling of technology nodes as discussed in the previous section.
A better circuit-design approach for tolerating wearout is through optimal sizing.
Specifically, for BTI, upsizing the transistors can compensate for the V_{th} increase.
EM effects are mainly addressed by design rules (e.g., increasing the metal width)
during the physical design phase. Sizing is a complex problem for solvers and,
as it is associated with multiple metrics, optimizing for one can hurt another.
Besides the area overhead, the increased transistor sizes contribute to increased
gate capacitances, thus increased dynamic power consumption, and it also increases
the leakage power. Similarly, increased metal width contributes to increased load
capacitance, which can increase the power consumption as well. Another design-
time method to address BTI degradation problems is during the technology mapping
of the logic synthesis. The idea is to balance the delays of system components

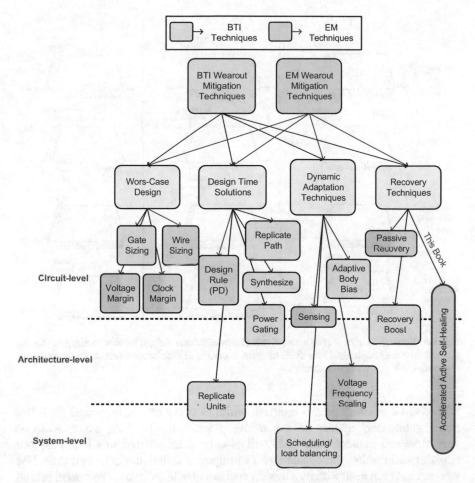

Fig. 1.4 Taxonomy of BTI and EM mitigation techniques. Note that "recovery boost" here refers to the existing recovery acceleration solutions which mainly focus on BTI wearout for SRAMs

(gates, paths, or even processor pipeline stages) by considering wearout so that the overall lifetime of the system can be optimized [31]. As this solution is strongly based on being able to predict wearout under dynamic conditions, it can lead to overestimation and inaccurate results in many cases, and this in turn leads to a low efficiency for design-time approaches. An alternative design-time solution has been adding redundant resources for wearout-critical components such as critical path at the circuit level [32] or cores/processing units at the architecture/system level [33]. The overall lifetime can then be improved by switching among redundant sources. Adding redundant elements can increase the area significantly, can also lead to performance overhead during switching, and it complicates the design process as well. Power gating has been a popular low-power design techniques that was originally used to lower the leakage power. It has been also adapted to help relieve

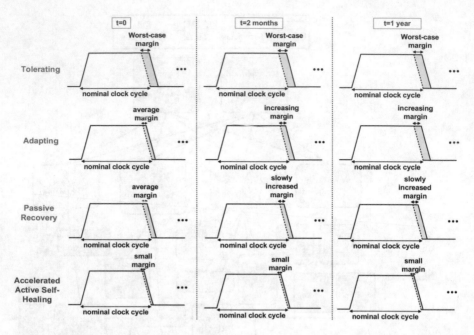

Fig. 1.5 Illustration of how each wearout mitigation technique assigns the design margins (timing margins in this example) and how these margins evolve with time under wearout; similar margins can be described as voltage headroom

BTI wearout as stress time is reduced, while recovery time is increased [34]. But power gating only enables passive recovery, which has been demonstrated to be very slow and unpredictable (this will also be demonstrated in Chap. 2 by our experimental results). Compared to BTI mitigation techniques at design time, EM wearout has been dealt with by a less diverse group of techniques. The reason behind this is that front-end designers (who design the architectures and IPs) have much less control over backend implementations, so physical designers usually need to take care of EM wearout by either upsizing the metal wire based on the design rules specified by the foundry or by adding more metal straps to compensate for the resistance increase caused by EM. This solution, although being accepted for many years, is known to be inefficient as it can lead to wasted routing resources and conservative design.

Compared to the static design-time solutions, adaptive post-silicon techniques appear to be more "economic" in terms of costs and margins by compensating wearout during run time. Previous works have proposed novel circuit- and architecture-level BTI [35] and EM sensors [36] to track and monitor wearout, then several *knobs* can be adjusted correspondingly. Such knobs can be the clock frequency [37], supply voltage [30], body bias [38], or a mix of all [39]. At the system level, BTI-aware scheduling was proposed to equalize the utilization of functional units in a microprocessor to improve its lifetime reliability [40]. Although the dynamic margins enabled by these solutions can guarantee that the circuit or system is functioning in the presence of wearout, wearout itself is still left unchecked. Under these conditions, the system can function but might run sluggish or burn more power gradually (illustrated as the "adapting" case in Fig. 1.5). Also due to the unique time-dependent nature of wearout, the irreversible components will accumulate as the system runs. In many cases, wearout sensors (the expected number for a future chip can reach as many as hundreds [35]) need to be ON for tracking over the entire lifetime, and this may add unacceptable tracking power overhead. A better solution would be to somehow reduce the actual wearout-induced variations by "repairing" them. Since BTI/EM are voltage/current dependent [41], one way is to reduce the voltage/current stress, thus to alleviate wearout during run time [30], but this approach could introduce big performance overhead. The second way is to take advantage of the recovery properties of BTI/EM by generating more idle time for passive recovery (system unstressed when not in use) [42]. Passive recovery is slow and unpredictable, but it might reduce some necessary design margins compared to the adaptive solutions, albeit at a very low rate (this has been illustrated as the "passive recovery" case in Fig. 1.5). Due to its unpredictability and slow rate passive recovery is sometimes even ignored when modeling wearout phenomena.

Based on the previous discussion, a solution that can fundamentally fix wearout would be clearly preferable to just adapting or compensating for its effects. The concept of recovery boost was firstly introduced in SRAMs for cache blocks to recover some of the NBTI effects through reverse bias [2, 40], but these ideas were mostly based on intuition due to lack of experimental results and good understanding of device-level recovery properties; also the implementation in these works cannot be applied to other logic and did not consider other wearout mechanisms such as EM. Different from previous wearout mitigation solutions, this book introduces a new effective direction that repairs both BTI and EM wearout in a more complete and fundamental way. We propose and demonstrate that wearout can be made *active* by *reversing* the directions of the stress (e.g., using positive V_{gs} instead of negative V_{gs} for NBTI, using reverse current instead of forward current for EM) and can be *accelerated* (e.g., by increasing the temperature). Based on actual hardware measurement results, these *accelerated active self-healing*[2] techniques will lead to significant recovery rate improvements, thus leading to much less pessimistic design guardbands (illustrated as the "accelerated active self-healing" case in Fig. 1.5). On top of these, we investigate the irreversible components of both wearout effects, and propose a set of biologically inspired solutions which can almost fully mitigate and avoid these components. On-chip implementations across the full system stack (circuit, architecture, and system) that enable and assist both BTI and EM accelerated self-healing will also be discussed in detail. As technology advances it introduces potential reliability concerns that are dominated by wearout,

---

[2]In this book, terms "accelerated active self-healing" and "accelerated and active recovery" are used interchangeably.

especially in emerging applications such as Internet of Things (IoT). To secure a resilient operations within the required lifetime, accelerated active self-healing can be a promising solution that takes advantage of recovery properties of the on-chip components (transistors or wires) and the intrinsic sleep behavior of the application itself.

1.3 Overview of This Book

In this book, we identify the problem that future electronic systems will suffer wearout issues from both technology and application aspects, and there hasn't been a single wearout mitigation solution that can serve as a "panacea" yet. The goal of the book is to discuss an effective (in terms of power, performance, and area) and orthogonal (that can be used together with other existing solutions) dimension of fixing wearout through accelerated and active recovery [43]. We present our research results on experimental validations [44–46], implementations [47–49], and applications [50, 51]. The main contributions of this book are summarized as follows:

1. *Providing new insights of BTI and EM recovery behaviors through comprehensive studies based on hardware measurements.* Most of the wearout experimental research in the past mainly looked into the stress phase behavior, while recovery for both EM and BTI hasn't been properly exploited yet. In particular, there are irreversible components for both wearout effects, and the explanation of this has been manifold due to disagreements about the physics phenomena, and lack of comprehensive experimental results. In this book, we discuss the recovery behavior based on measurement on actual hardware (FPGAs for BTI, fabricated test chip metal wires for EM). Each set of measurements has been designed by considering different combinations of recovery conditions and lasts for more than 3 days. We also explore the boundary between the reversible and irreversible components of BTI and EM. The experimental results provide many new insights on recovery, such as *frequency dependency, accelerated and active recovery*, and *long-term recovery vs. short-term recovery differences*. These insights contribute as experimental evidence and validation results that the reliability community can use to develop better and more accurate BTI and EM models.

2. *Demonstrating that accelerated active self-healing is an effective solution for fixing wearout.* Accelerated active self-healing is essentially a "reverse" wearout process where several knobs are tuned during recovery to accelerate the process. We demonstrate that accelerated active self-healing can lead to significant recovery rate improvements (e.g., 72.4% of the wearout is recovered within only 1/4 of the stress time for BTI; 70% of EM wearout can be recovered within 1/5 of the stress time). Our study demonstrates that there is still a lingering permanent component which is irreversible even under accelerated and active recovery conditions, and we explore the boundary between the reversible and irreversible

wearout components physically and experimentally. By studying the frequency dependent behaviors of BTI and EM wearout and recovery, we demonstrate that the above boundary is actually "soft" and can even be controllable, and this leads to a biologically inspired sleep-when-getting-tired strategy that keeps the circuit active only during the reversible phase of wearout and puts it in active recovery mode before the irreversible wearout kicks in, thus the irreversible wearout can become almost unobservable even under the accelerated stress cases. The accelerated active self-healing solutions, together with scheduled explicit accelerated self-healing periods ahead of any sign of permanent stress, is simpler to implement on chip and means that the system will operate for a longer time in a "refreshed" mode, thus leading to better performance, and has better cumulative metrics (e.g., average performance) as well.

3. *Developing the infrastructure for enabling future research in resilient system design.* To enable the implementations of accelerated self-healing techniques and fully utilize the unique recovery behaviors for BTI and EM wearout, we present a full set of potential implementation solutions at the circuit, architecture, and system levels. These solutions cover all aspects, from recovery-driven design methodology, novel sensing techniques for monitoring both wearout and recovery, novel power gating structures, and recovery assist circuit for enabling multiple recovery modes. Some of these circuit solutions have also been successfully demonstrated with fabricated test chips. Since single-layer recovery solutions are not the most cost-friendly, we also discuss the implications of implementing accelerated self-healing at different levels of a system hierarchy. The recovery circuit IP components are portable and flow friendly, and thus can serve as the infrastructures for future research in this direction. Combining all these techniques enables a true accelerated self-healing system that benefits from the full-recovery capabilities.

4. *Filling the knowledge gap between planar transistors and FinFETs and raising the awareness of FinFET wearout in emerging applications such as IoT domain.* As FinFET has become a mainstream technology node for many high-performance computing chips, it is also in the process of being adopted for low-cost embedded systems; in this book, we present a comprehensive technology study across multiple nodes ranging from planar, FD-SOI to FinFET based on both foundry-provided models and predictive models. The study explores new design challenges and new insights since the advent of FinFET technology. It can thus be used as an educational material and design guide for circuit designers who design with FinFETs. At advanced technology nodes, wearout has become more pronounced, and this can especially be critical in the context of IoT in which some applications require very reliable operations (zero-error tolerance) spanning a much longer lifetime (e.g., >50 years). The book looks into this aspect as well by studying how transistor wearout issues impact different categories of IoT applications with the foundry-provided wearout models. We conclude that wearout plays a key role in determining the overall IoT system lifetime and cannot be ignored. Through this study, we hope to raise the awareness of the impact of wearout in several critical IoT applications and encourage designers to take action in the first place.

The rest of the book will focus on three aspects of accelerated and active self-healing. Part II (including Chaps. 2 and 3) presents the experimental results for both BTI and EM recovery. Part III (including Chaps. 4 and 5) discusses the on-chip implementations that enable the accelerated and active recovery across the system hierarchy. In Part IV (Chap. 6), we provide technology studies on FinFETs vs. planar nodes, and we also explore the impact of wearout on IoT applications. Part V, which includes Chap. 7, concludes the book and discusses the future research directions that can be enabled by materials presented in this book.

References

1. Muhammad Shafique, Siddharth Garg, Jörg Henkel, and Diana Marculescu. The eda challenges in the dark silicon era: Temperature, reliability, and variability perspectives. In *Proceedings of the 51st Annual Design Automation Conference*, pages 1–6. ACM, 2014.
2. Jeonghee Shin, Victor Zyuban, Pradip Bose, and Timothy M Pinkston. A proactive wearout recovery approach for exploiting microarchitectural redundancy to extend cache sram lifetime. In *ACM SIGARCH Computer Architecture News*, volume 36, pages 353–362. IEEE Computer Society, 2008.
3. Robert Baumann. Soft errors in advanced computer systems. *IEEE Design & Test of Computers*, 22(3):258–266, 2005.
4. ITRS Report. http://www.itrs2.net/itrs-reports.html.
5. James H Stathis, M Wang, RG Southwick, EY Wu, BP Linder, EG Liniger, G Bonilla, and H Kothari. Reliability challenges for the 10nm node and beyond. In *Electron Devices Meeting (IEDM), 2014 IEEE International*, pages 20–6. IEEE, 2014.
6. Rob Aitken, Ethan H Cannon, Mondira Pant, and Mehdi B Tahoori. Resiliency challenges in sub-10nm technologies. In *VLSI Test Symposium (VTS), 2015 IEEE 33rd*, pages 1–4. IEEE, 2015.
7. Fabian Oboril and Mehdi B Tahoori. Cross-layer approaches for an aging-aware design of nanoscale microprocessors: Dissertation summary: IEEE TTTC E.J. McCluskey doctoral thesis award competition finalist. In *Test Conference (ITC), 2015 IEEE International*, pages 1–10. IEEE, 2015.
8. Shekhar Borkar. *Private communication*, 2018.
9. SM Ramey, C Prasad, and A Rahman. Technology scaling implications for BTI reliability. *Microelectronics Reliability*, 82:42–50, 2018.
10. Souvik Mahapatra. *Fundamentals of Bias Temperature Instability in MOS Transistors*. Springer, 2016.
11. Jacopo Franco, Salvatore Graziano, Ben Kaczer, Felice Crupi, L-Å Ragnarsson, Tibor Grasser, and Guido Groeseneken. Bti reliability of ultra-thin eot mosfets for sub-threshold logic. *Microelectronics Reliability*, 52(9):1932–1935, 2012.
12. Yu Cao, Jyothi Velamala, Ketul Sutaria, Mike Shuo-Wei Chen, Jonathan Ahlbin, Ivan Sanchez Esqueda, Michael Bajura, and Michael Fritze. Cross-layer modeling and simulation of circuit reliability. *Computer-Aided Design of Integrated Circuits and Systems, IEEE Transactions on*, 33(1):8–23, 2014.
13. Xin Huang, Valeriy Sukharev, Taeyoung Kim, and Sheldon X-D Tan. Dynamic electromigration modeling for transient stress evolution and recovery under time-dependent current and temperature stressing. *Integration, the VLSI Journal*, 2016.
14. C Prasad, S Ramey, and L Jiang. Self-heating in advanced cmos technologies. In *Reliability Physics Symposium (IRPS), 2017 IEEE International*, pages 6A–4. IEEE, 2017.
15. Brian Bailey. Chip aging becomes design problem. *Semiconductor Engineering*, 2018.

16. Wenping Wang, Shengqi Yang, Sarvesh Bhardwaj, Rakesh Vattikonda, Sarma Vrudhula, Frank Liu, and Yu Cao. The impact of nbti on the performance of combinational and sequential circuits. In *Proceedings of the 44th annual Design Automation Conference*, pages 364–369. ACM, 2007.
17. Weisong Shi and Schahram Dustdar. The promise of edge computing. *Computer*, 49(5):78–81, 2016.
18. Jan M Rabaey, Anantha P Chandrakasan, and Borivoje Nikolic. *Digital integrated circuits*, volume 2. Prentice hall Englewood Cliffs, 2002.
19. Ed Sperling. Chip Aging Accelerates. *Semiconductor Engineering*, 2018.
20. Massimo Alioto. Enabling the Internet of Things: From Integrated Circuits to Integrated Systems. Springer, 2017.
21. Jörg Henkel, Lars Bauer, Nikil Dutt, Puneet Gupta, Sani Nassif, Muhammad Shafique, Mehdi Tahoori, and Norbert Wehn. Reliable on-chip systems in the nano-era: lessons learnt and future trends. In *Proceedings of the 50th Annual Design Automation Conference*, page 99. ACM, 2013.
22. Bipul C Paul, Kunhyuk Kang, Haldun Kufluoglu, Muhammad Alam, Kaushik Roy, et al. Impact of nbti on the temporal performance degradation of digital circuits. *Electron Device Letters, IEEE*, 26(8):560–562, 2005.
23. Hyejeong Hong, Jaeil Lim, Hyunyul Lim, and Sungho Kang. Lifetime reliability enhancement of microprocessors: Mitigating the impact of negative bias temperature instability. *ACM Computing Surveys (CSUR)*, 48(1):9, 2015.
24. Runsheng Wang, Pengpeng Ren, Changze Liu, Shaofeng Guo, and Ru Huang. Understanding nbti-induced dynamic variability in the nano-reliability era: From devices to circuits. In *Physical and Failure Analysis of Integrated Circuits (IPFA), 2015 IEEE 22nd International Symposium on the*, pages 119–121. IEEE, 2015.
25. Kerry Bernstein, David J Frank, Anne E Gattiker, Wilfried Haensch, Brian L Ji, Sani R Nassif, Edward J Nowak, Dale J Pearson, and Norman J Rohrer. High-performance cmos variability in the 65-nm regime and beyond. *IBM journal of research and development*, 50(4.5):433–449, 2006.
26. Sheldon X-D Tan, Hussam Amrouch, Taeyoung Kim, Zeyu Sun, Chase Cook, and Jörg Henkel. Recent advances in em and bti induced reliability modeling, analysis and optimization. *Integration, the VLSI Journal*, 2017.
27. D. C. Sekar et al. Electromigration Resistant Power Delivery Systems. *IEEE Electron Device Letters*, 28(8):767–769, Aug 2007.
28. Navid Khoshavi, Rizwan A Ashraf, Ronald F DeMara, Saman Kiamehr, Fabian Oboril, and Mehdi B Tahoori. Contemporary CMOS aging mitigation techniques: Survey, taxonomy, and methods. *Integration, the VLSI Journal*, 59:10–22, 2017.
29. Kunhyuk Kang, Saakshi Gangwal, Sang Phill Park, and Kaushik Roy. NBTI induced performance degradation in logic and memory circuits: how effectively can we approach a reliability solution? In *Proceedings of the 2008 Asia and South Pacific Design Automation Conference*, pages 726–731. IEEE Computer Society Press, 2008.
30. Lide Zhang and Robert P Dick. Scheduled voltage scaling for increasing lifetime in the presence of nbti. In *Proceedings of the Asia and South Pacific Design Automation Conference*, pages 492–497. IEEE Press, 2009.
31. Saman Kiamehr, Farshad Firouzi, Mojtaba Ebrahimi, and Mehdi B Tahoori. Aging-aware standard cell library design. In *Proceedings of the conference on Design, Automation & Test in Europe*, page 261. European Design and Automation Association, 2014.
32. Rizwan A Ashraf, Navid Khoshavi, Ahmad Alzahrani, Ronald F DeMara, Saman Kiamehr, and Mehdi B Tahoori. Area-energy tradeoffs of logic wear-leveling for bti-induced aging. In *Proceedings of the ACM International Conference on Computing Frontiers*, pages 37–44. ACM, 2016.
33. Jayanth Srinivasan, Sarita V Adve, Pradip Bose, and Jude A Rivers. Exploiting structural duplication for lifetime reliability enhancement. In *Computer Architecture, 2005. ISCA'05. Proceedings. 32nd International Symposium on*, pages 520–531. IEEE, 2005.

34. Navid Khoshavi, Rizwan A Ashraf, and Ronald F DeMara. Applicability of power-gating strategies for aging mitigation of CMOS logic paths. In *Circuits and Systems (MWSCAS), 2014 IEEE 57th International Midwest Symposium on*, pages 929–932. IEEE, 2014.
35. S Sarma, N Dutt, N Venkatasubramanian, A Nicolau, and P Gupta. Cyberphysical system-on-chip (cpsoc): Sensor actuator rich self-aware computational platform. *University of California Irvine, Tech. Rep. CECS TR-13-06*, 2013.
36. Kai He, Xin Huang, and Sheldon X-D Tan. Em-based on-chip aging sensor for detection and prevention of counterfeit and recycled ics. In *Computer-Aided Design (ICCAD), 2015 IEEE/ACM International Conference on*, pages 146–151. IEEE, 2015.
37. Fabian Oboril and Mehdi B Tahoori. Reducing wearout in embedded processors using proactive fine-grain dynamic runtime adaptation. In *Test Symposium (ETS), 2012 17th IEEE European*, pages 1–6. IEEE, 2012.
38. Zhenyu Qi and Mircea R Stan. Nbti resilient circuits using adaptive body biasing. In *Proceedings of the 18th ACM Great Lakes symposium on VLSI*, pages 285–290. ACM, 2008.
39. Evelyn Mintarno et al. Self-tuning for maximized lifetime energy-efficiency in the presence of circuit aging. *IEEE TCAD*, 30(5):760–773, 2011.
40. Taniya Siddiqua and Sudhanva Gurumurthi. Nbti-aware dynamic instruction scheduling. In *Proceedings of the 5th Workshop on Silicon Errors in Logic-System Effects*. Citeseer, 2009.
41. Jyothi Bhaskarr Velamala, Ketul Sutaria, Takashi Sato, and Yu Cao. Physics matters: statistical aging prediction under trapping/detrapping. In *Proceedings of the 49th Annual Design Automation Conference*, pages 139–144. ACM, 2012.
42. Saket Gupta and Sachin S Sapatnekar. Employing circadian rhythms to enhance power and reliability. *ACM Transactions on Design Automation of Electronic Systems (TODAES)*, 18(3):38, 2013.
43. Xinfei Guo. Towards Wearout-Aware and Accelerated Self-Healing Digital Systems. PhD Thesis, University of Virginia, 2018.
44. Xinfei Guo, Wayne Burleson, and Mircea Stan. Modeling and experimental demonstration of accelerated self-healing techniques. In *Design Automation Conference (DAC), 2014 51st ACM/EDAC/IEEE*, pages 1–6. IEEE, 2014.
45. Xinfei Guo and Mircea R Stan. Work hard, sleep well-avoid irreversible ic wearout with proactive rejuvenation. In *Design Automation Conference (ASP-DAC), 2016 21st Asia and South Pacific*, pages 649–654. IEEE, 2016.
46. Xinfei Guo and Mircea R Stan. Enabling Wearout-Immune BEOL and FEOL with Active Rejuvenation. In *IEEE/ACM Workshop on Variability Modeling and Characterization (VMC)*, 2017.
47. Xinfei Guo and Mircea R Stan. MCPENS: Multiple-Critical-Path Embeddable NBTI Sensors for Dynamic Wearout Management. In *Workshop on Silicon Errors in Logic-System Effects (SELSE-11)*. IEEE, 2015.
48. Xinfei Guo and Mircea R Stan. Implications of accelerated self-healing as a key design knob for cross-layer resilience. *Integration, the VLSI Journal*, 56:167–180, 2017.
49. Xinfei Guo and Mircea R Stan. Deep Healing: Ease the BTI and EM Wearout Crisis by Activating Recovery. In *Dependable Systems and Networks Workshop (DSN-W), 2017 47th Annual IEEE/IFIP International Conference on*, pages 184–191. IEEE, 2017.
50. Xinfei Guo, Vaibhav Verma, Patricia Gonzalez-Guerrero, Sergiu Mosanu, and Mircea R Stan. Back to the Future: Digital Circuit Design in the FinFET Era. *Journal of Low Power Electronics*, 13(3):338–355, 2017.
51. Xinfei Guo, Vaibhav Verma, Patricia Gonzalez-Guerrero, and Mircea R Stan. When "things" get older - Exploring Circuit Aging in IoT Applications. In *Quality Electronic Design (ISQED), International Symposium on*. IEEE, 2018.

Part II
Experimental Validations

Chapter 2
Accelerated and Active Self-healing Techniques for BTI Wearout

2.1 Overview

Bias temperature instability (BTI) is one of the most dominant wearout mechanisms that occur to transistors, especially in advanced technology nodes [1]. BTI increases the absolute value of the threshold voltage (V_{th}) and reduces the mobility (μ) of transistors over time under voltage stress, thus increasing the circuit delay and necessary time margins [2, 3]. As shown in Fig. 2.1, negative bias temperature instability (NBTI) occurs under negative stress conditions and affects PMOS transistors. Similarly, positive bias temperature instability (PBTI) affects NMOS transistors under positive stress voltage. Depending on the bias condition of the gate, there are two phases for BTI. The *stress* phase happens when the gate is under voltage stress ($V_{gs} < 0$ for PMOS, $V_{gs} > 0$ for NMOS), and the *passive recovery* phase happens when transistors are in OFF state, where voltage stress is "released" ($V_{gs} = 0$). While passive recovery has been accepted as being slow and unpredictable, thus cannot be reliably used to reduce margins and it is sometimes even ignored when modeling wearout phenomena and estimating the necessary guardbands. Different from previous recovery solutions, in this chapter, we present experimental results that demonstrate that BTI recovery can actually be made *active* by reversing the direction of the voltage stress (e.g., using positive instead of negative V_{gs} to reverse NBTI), and can be accelerated (e.g., by increasing the temperature), thus leading to accelerated active self-healing. Based on actual hardware measurement results on commercial FPGAs in a 40 nm technology node, these accelerated self-healing techniques can lead to significant recovery rate improvements (e.g., we demonstrate a case where 72.4% of the wearout is recovered within only 1/4 of the stress time). We also find that even under the accelerated active recovery case, there are still components of BTI wearout that are irreversible. By studying the frequency dependent behavior of BTI wearout and recovery, we demonstrate that the boundary between reversible and irreversible is actually "soft" and can even be controllable; this leads to a biologically inspired sleep-when-getting-tired strategy that *keeps the*

© Springer Nature Switzerland AG 2020
X. Guo, Mircea R. Stan, *Circadian Rhythms for Future Resilient Electronic Systems*, https://doi.org/10.1007/978-3-030-20051-0_2

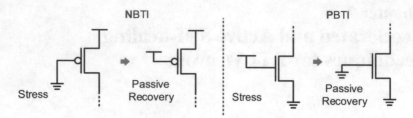

Fig. 2.1 BTI stress and passive recovery: NBTI happens in PMOS transistors; PBTI happens in NMOS transistors. For both mechanisms, BTI starts recovering when transistors are turned OFF, but this passive recovery period is slow and unpredictable

circuit active only during the reversible phase of wearout before the irreversible wearout kicks in, then followed by in-time accelerated active recovery, thus the overall irreversible wearout becomes almost unobservable even under accelerated stress cases. The proposed solutions, with proactively scheduled explicit accelerated active self-healing periods ahead of any sign of stress, will be simpler to implement on chip and the system can run mostly in a "refreshed" mode, thus leading to less necessary BTI-determined margins, better performance, and has better cumulative metrics (e.g., average performance) as well. Details of the experimental setup, measurement results, and analytic modeling of the accelerated active self-healing solutions are presented in this chapter.

2.2 BTI Wearout and Recovery Basics

The understanding of the mechanisms behind BTI has been somewhat controversial and there is still no full consensus about the physics behind it [4, 5]. There are two main theories that try to explain BTI, and they are illustrated in Fig. 2.2. The first one is called the "reaction–diffusion (RD)" theory [6] and is shown in Fig. 2.2a. According to this theory, BTI has been attributed mainly to interface trap generation. For example, for the NBTI case (PMOS), when the transistor is ON, the voltage stress across gate and source could potentially break the covalent bond of Si–H at interface—this process is called *reaction*. The separated hydrogen atoms combine to form H_2, which diffuses towards the gate of the transistor. These broken Si–H bonds generate positively charged traps for holes and lead to transistor parametric shift, such as increased threshold voltage V_{th}. When the PMOS switches off (i.e., $V_{gs} = 0$), and stress is removed, the recovery phase starts, where holes are not present in the channel and thus, no new interface traps are generated; instead, H diffuses back and anneals the broken Si–H bond. As a result, the number of interface traps is reduced during this phase and the NBTI degradation is recovered (passively).

While the RD based theory can predict BTI for constant stress, recent advances in fast on-chip BTI measurement have explored several dynamic BTI behaviors that cannot be explained and are inconsistent with what has been modeled by RD

Fig. 2.2 Two types of
explanations for BTI wearout.
(**a**) BTI reaction–diffusion
(RD) mechanism. (**b**) BTI
trapping–detrapping (TD)
mechanism

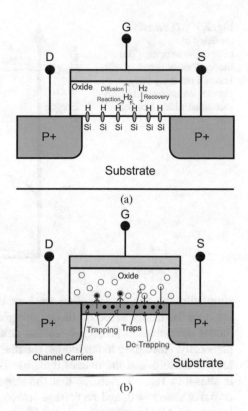

(a)

(b)

theory [7]. This led to an alternative way to explain BTI using the trapping/detrapping (TD) mechanism, which has been validated against silicon and became widely used in the community [2, 8–10]. As shown in Fig. 2.2b, when the PMOS transistor is ON, the trap energy (relative to the Fermi energy level) is modulated. If the trap gains sufficient energy, it may capture a charge carrier, thus reducing the number of available carriers in the channel, the charged trap state modulates the V_{th} and acts as a scattering source, reducing the effective mobility. This is called the *trapping* process. Similar to what has been captured by the RD theory, if the transistor is OFF and in passive recovery phase, some of the interface traps are being slowly annealed (shown as *detrapping* process in the figure), and the number of occupied traps reaches a new equilibrium and results in partial recovery. Although only NBTI (PMOS) has been dominant and usually the effect of PBTI (for NMOS) has been negligible in older technologies, PBTI has become equally important with the introduction of high-k and metal gates [11–13]. Since the degradation effect of PBTI is similar to NBTI, the PBTI effect can be modeled similar to the NBTI effect [13].

There have been several recent modeling frameworks which incorporate both the RD and TD theory [14, 15]. These works have shown that the most dominant component of BTI is due to hole trapping and interface trap generation. Thus, for

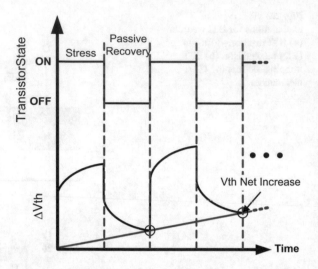

Fig. 2.3 BTI behaviors modeled by trapping/detrapping (TD) theory. Passive recovery still leaves a net ΔV_{th} that is almost permanent and hardly recovered within a reasonable time

simplicity, the chosen modeling work in this chapter and throughout the rest of the book is mainly based on the trapping/detrapping theory. The device-level TD model developed in [16] captures the details of the stress and passive recovery physically. According to the model, the threshold voltage of the transistor increases logarithmically, and the overall dynamic BTI behaviors can follow a pattern that is shown in Fig. 2.3. Assume that the single transistor is turned on (stress period starts) at time $t = 0$, and no voltage stress is applied before. The threshold voltage increases (ΔV_{th}) until time t_{st} is modeled as:

$$\Delta V_{th}(t_{st}) = \phi_{st} \left(A + log(1 + Ct_{st})\right) \tag{2.1}$$

If a recovery interval of t_{rec} follows the stress phase, the total threshold voltage shift $\Delta V_{th}(t_{st} + t_{rec})$ in the end is equal to:

$$\Delta V_{th}(t_{st} + t_{rec}) = \phi_{rec} \left(A + log(1 + Ct_{rec})\right) + \Delta V_{th}(t_{st})$$
$$\times \left(1 - \frac{A + log(1 + Ct_{rec})}{A + log(1 + C(t_{st} + t_{rec}))}\right) \tag{2.2}$$

$$\phi \sim K exp\left(\frac{-E}{kT}\right) exp\left(\frac{BV_{dds/r}}{kT t_{ox}}\right) \tag{2.3}$$

where A, B, C are (approximately) constant across the same technology node, K is the fitting parameter, k is Boltzmann's constant, T is temperature, E is activation energy, t_{ox} is the oxide thickness, and V_{dds} and V_{ddr} are the supply voltages under stress and recovery, respectively. More details of the model can be found in [16].

The device-level BTI model discussed above indicates the strong (exponential) dependence of the threshold voltage shift on voltage and temperature during both stress and recovery period. It serves as the basis for our circuit and higher level accelerated active self-healing modeling framework. All parameters and their values used in the model are extracted based on measurements which will be discussed in detail in the following sections.

2.3 Prior Work on BTI Recovery

The partial recovery property of BTI has been utilized in many works to improve the lifetime and other metrics (e.g., performance) of the digital systems. Several approaches [17–21] were proposed to rebalance the signal probabilities for logic or SRAMs to maximize the passive recovery time at the circuit level. An alternative method was introduced to adaptively tune the performance according to the degree of BTI wearout so that certain blocks could start the passive recovery phase earlier [22, 23]. Novel schemes were proposed to exploit the idle time of busy functional units for out-of-order processors [24] and superscalar architectures [25]. A dynamic routing algorithm was proposed to adapt to the wearout-induced degradation in heterogeneous NoCs [26]. Since passive recovery is much slower than the wearout process [27], *wearout gating* [28] was proposed—the idea is to add a coupling transistor to a regular power gating structure so that the virtual supply and the logic high/low supply are equalized; in this way, the voltage across the logic block is zero and transistors experience zero stress. The goal of this method was still to release the stress completely for passive recovery. To further boost the recovery process, *intense recovery* for logic [29] and SRAM arrays [30, 31] was then proposed, the idea was to raise the gate voltage of a chain of logic or memory cell in order to put PMOS devices into the recovery enhancement mode in which the voltage across the transistor is reverse bias mode with full range of V_{dd}. This method can potentially be harmful to device reliability since it can lead to transistor breakdown under a high voltage, it also incurs very high power routing costs and design complexity. A power napping scheme was proposed to help the recovery of NBTI and PBTI [32]. But all of these previous work focused on only SRAM cell designs or architectural level implementations, it was still unclear how much benefit recovery boost could achieve due to lack of experimental data and models. In addition, these solutions still leave the irreversible wearout unchecked. Wafer level and transistor level experiments and theory [33–35], together with the device-level model discussed in the previous section, indicated that BTI recovery highly depends on temperature; these works provide physical evidence for the accelerated self-healing techniques detailed in this book. Several recent works [15, 36, 37] have studied the irreversible component of wearout at the device level; however, these works focus only on demonstrating and modeling the permanent component, thus a solution that can fundamentally repair the irreversible wearout is still missing in the field. The solutions presented in this chapter differ from the previous recovery and recovery boosting work in several aspects. First, we demonstrate that both high temperature and negative voltage can accelerate recovery with actual hardware measurements on commercial FPGAs and also develop the analytic model for them. Second, we propose a biologically inspired strategy that runs the system in such a way that the irreversible wearout can be fully avoided so that the system remains almost "fresh" all the time. Lastly, the self-healing solutions can be effectively applied to all logic blocks, instead of just SRAM arrays. The notion of accelerated self-healing can be implemented on chip across the system hierarchy and can be introduced as a key design knob for cross-layer resilience.

2.4   BTI Self-healing

2.4.1   Accelerate and Activate BTI Recovery

In this book, we postulate that the future electronic systems can use sleep time as an active recovery period essential for their overall performance. We demonstrate that during sleep, several *accelerated self-healing* solutions can be implemented to deeply rejuvenate the circuit, and they are shown in Fig. 2.4 (we illustrate only NBTI recovery as an example). First, BTI recovery can be made *active* by *deeply* "turning off" the transistor via a negative voltage across the source and gate. Second, high temperature can increase the kinetic energy for the charge carriers, thus *accelerating* the recovery. The third case is when active recovery can be further accelerated through the joint effort of both negative voltage and high temperature. From the physics perspective, the self-healing techniques reverse the direction of BTI wearout and increase the rate and level of recovery. The applied negative voltage (active recovery) across the transistors activates the detrapping process by pushing trapped carriers back to their original states, while high temperature can further assist the healing process by exciting the carriers. Overall, two solutions work together to achieve the highest possible rate and level of recovery.

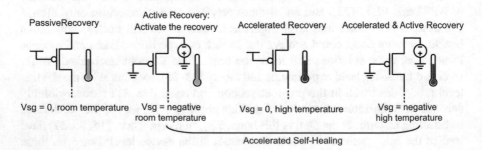

Fig. 2.4 Accelerated active self-healing solutions for NBTI, similar solutions can be applied to PBTI

2.4.2 Gate-Level Analytical Model for Accelerated Active Self-healing

To model the performance degradation and recovery due to wearout and accelerated self-healing, a gate-level analytical model that is based on the device model described in (2.1)–(2.3) is developed in this book. The model serves as a connection between circuit metrics (such as delay t_d) and device-level parameter shift (mainly V_{th} increase). Based on transistor theory [38], the propagation delay of a digital gate can be approximated as:

$$t_d \sim \frac{C_L V_{dd}}{I_d} \propto \frac{C_L V_{dd}}{V_{dd} - V_{th}} \tag{2.4}$$

where C_L is the output capacitance of the gate. The change in gate delay when V_{th} is subject to change is:

$$\Delta t_d \sim \frac{\Delta V_{th}}{V_{dd} - V_{th}} \cdot t_{d0} \tag{2.5}$$

where t_{d0} is the original delay of the gate without any V_{th} shift. Combine Eqs. (2.1), (2.3), and (2.5), the total delay increase after t_{st} can be expressed as:

$$\Delta t_d(t_{st}) = \beta_{st} exp \left(\frac{-E}{kT} \right) exp \left(\frac{BV_{dds}}{kT t_{ox}} \right) (A + log(1 + Ct_{st})) \tag{2.6}$$

where β_{st}, A, B, and C are fitting parameters and can be extracted from measurement results. During the accelerated active self-healing phase, based on the recovery phase equation of the device model, we combine Eqs. (2.2), (2.3), and (2.5), and the delay change after sleep period t_{rec} becomes:

$$\Delta t_d(t_{st} + t_{rec}) = \frac{\phi_{rec}}{V_{dds}} (A + log(1 + Ct_{rec})) + \Delta t_d(t_{st})$$

$$\times \left(1 - \frac{A + log(1 + Ct_{rec})}{A + log(1 + C(t_{st} + t_{rec}))} \right) \tag{2.7}$$

Assume that the ratio of operation time to active sleep time of the system is α, the total time (stress time + sleep time) is t_{total}, based on (2.7), the overall delay increase will be:

$$\Delta t_d(t_{total}) = \phi_{acce} \left(A + log \left(1 + C \frac{t_{total}}{1 + \alpha} \right) \right) + \Delta t_d \left(\frac{\alpha t_{total}}{1 + \alpha} \right)$$

$$\times \left(1 - \frac{A + log \left(1 + C \frac{t_{total}}{1 + \alpha} \right)}{A + log(1 + Ct_{total})} \right) \tag{2.8}$$

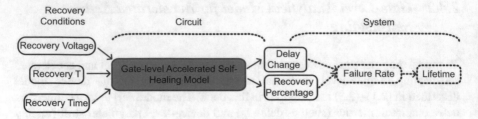

Fig. 2.5 Illustration of a potential use case for gate-level accelerated self-healing model. This flow can be used for evaluating how recovery conditions can further affect the circuit metrics and system lifetime

$$\phi_{acce} \sim \frac{K}{V_{ddr}} exp \left(\frac{-E}{kT_{acce}} \right) exp \left(\frac{B V_{ddr}}{kT_{acce} t_{ox}} \right) \tag{2.9}$$

Several observations can be made based on the gate-level accelerated self-healing model. First, the exponential dependency of the total delay increase on recovery temperature and voltage shows that by increasing T_{acc} and decreasing V_{ddr}, the first component in Eq. (2.8) can decrease significantly. The second observation is that the active vs. sleep ratio α also affects the overall delay change. The final observation is that the recovered part $(\Delta t_d(t_{total}) - \Delta t_d(t_{str}))$ highly depends on the previous stress history $(\Delta t_d(t_{str}))$.

This model captures the circuit-level metric change due to wearout and accelerated and active recovery. It can potentially be integrated into higher level abstract models for exploring the accelerated self-healing design space, and one such use case is illustrated in Fig. 2.5. The delay change due to wearout can lead to timing violations at the circuit level, and this could lead to failures at the system level. Based on the recovery conditions, the gate-level model is able to estimate the "recovered delay" which can potentially be used to predict the reduction of failure rate and extension of the overall system lifetime. Several other use cases of this model will be detailed in the following sections.

2.5 Experimental Setup

2.5.1 Test Platform

FPGA vendors have been aggressive recently in adopting advanced technology nodes; this makes FPGAs more susceptible to wearout that can lead to frequency degradation [39]. Due to their "bleeding edge" technologies, reconfigurability, and regular structure, FPGAs are an ideal test platform for wearout research [40, 41]. In the experiments presented in this book, we picked 2-input look-up-table (LUT)-based commercial FPGAs [42] fabricated in a 40 nm technology to demonstrate

Fig. 2.6 FPGA test platform architecture

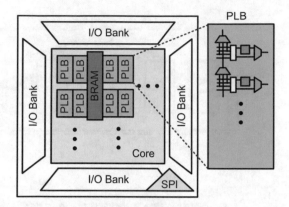

experimentally the introduced accelerated self-healing techniques described in the previous section. Figure 2.6 shows the architecture of the FPGA. Basic components of FPGAs include the I/Os and the core architecture; we use only the core architecture for testing wearout in the experiment. The core includes 1280 logic cells (LUT + flip-flops), which are grouped in programmable logic blocks (PLB) which can be programmed to perform logic and arithmetic functions. Each PLB consists of eight interconnected logic cells as shown in the figure. The chip has a SPI port that supports programming and configuration of the device using the standard FPGA synthesize and place and route (PnR) flow.

2.5.2 Test Configuration

In the gate-level model discussed in Sect. 2.4.2, the delay change is used as the metric to capture the effect of wearout and recovery, and the same metric is also employed in the experimental part. We choose a ring oscillator (RO) structure which is widely used as a test platform [43] to measure the delay of the circuit under test (CUT) to capture the delay change. Figure 2.7a shows the test configuration, which is a modified LUT-based ring oscillator based on a design proposed in [44]. It consists of 75 inverters implemented in LUTs and a 16-bit counter to capture the output frequency of the ring oscillator. Enable signal En is used to switch between AC stress (switching) and DC (constant) stress mode. Figure 2.7b shows how the test structure is mapped on the FPGA fabric. The CUT is kept under voltage stress and it is enabled only every 20 min for frequency recording. The oscillation frequency f_{osc} can be calculated as:

$$t_d = \frac{1}{2f_{osc}} = \frac{1}{4f_{ref}C_{out}} \tag{2.10}$$

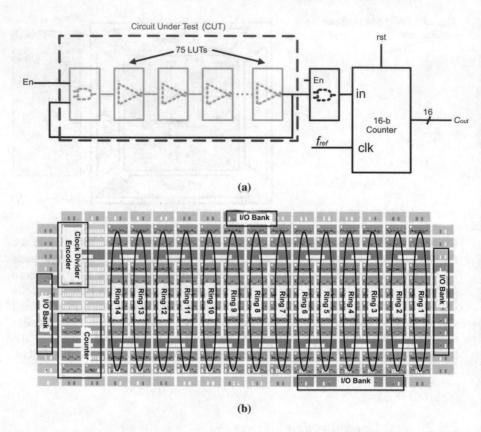

Fig. 2.7 BTI test configurations on FPGAs. (**a**) BTI test structure: a 75-stage ring oscillator mapping to the look-up-table (LUT) structure on FPGAs, the counter in the end is used to digitize the frequency differences. (**b**) A floorplan showing how BTI test structure is mapped on FPGA fabric: green squares represent utilized logic, red represents interconnect, I/Os interface with the host mother board and programmer through flat cable

where f_{ref} is the frequency of the reference clock. To pick this frequency, the CUT is placed at different locations on the FPGA shown as different indices in Fig. 2.7b (ring X), and a diagnostic program is run. The output of the counter is read from a certain time range that has stable values. Environmental factors and the voltage supply are kept constant from one reading to another; when $f_{ref} = 500$ Hz, the variation of the counter output is within ± 5 and $\pm 0.0001\%$ in terms of the corresponding RO frequency variation which we consider acceptable.

To sample the stress and recovery data from the FPGA chip in a fast and efficient manner, an automatic test flow is designed and is shown in Fig. 2.8. The FPGA chip interfaces with the PC through an evaluation board (AT91SAM7SE-EK) developed by Atmel [45]. The micro-controller unit (MCU) on the board can be programmed in C and serves as a "signal generator" for the control signals in the BTI test

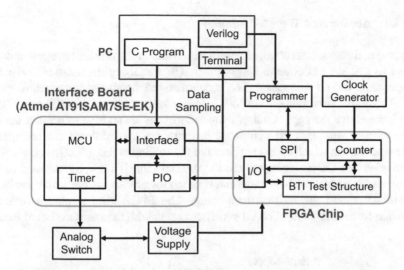

Fig. 2.8 BTI test flow: FPGA chip communicates with the computer through an Atmel evaluation board, on which a micro-controller can be programmed in C. The inherent timer in the MCU can be programmed to control how long the chip is stressed or recovered

structures on FPGA chip. The timer/counter in MCU is employed to control how long the FPGA chip is stressed and recovered, it also controls the analog switch which can pass or cut the supply voltage for the test chip. On the FPGA side, the RTL description of the BTI test structure (shown in Fig. 2.7a) is loaded into the FPGA through a programmer and serial peripheral interface (SPI). The reference clock for the counter is given by an external clock generator. The counter outputs are read out and saved to the PC through the mother board interface. By using this test methodology, the total data sampling overhead is less than 3 s, which should have a negligible impact on overall BTI wearout and recovery behaviors.

2.5.3 Test Conditions

2.5.3.1 Stress and Recovery "Knobs"

In Sect. 2.4, we have introduced three ways to accelerate the recovery during sleep, one is active recovery through small negative voltage, another is the high temperature, and the third is the combination of the first two. Thus the main "knobs" tuned in measurements are the voltage, time, temperature, switching activity and α, the ratio of stress (active) and recovery (sleep or rejuvenation) time. Two stress modes are considered—AC stress and DC stress (constant stress). DC stress refers to the case where the input of the test structure is always static and doesn't switch, and AC stress is when the input is switching with 50% duty cycle (En signal shown in Fig. 2.7a can switch between modes).

2.5.3.2 Accelerated Test Methodology

In our experiments, both elevated temperatures and high voltages are applied during stress so that we can observe a larger than 1% frequency degradation under high temperature for all test cases. The recommended operating temperature of the FPGAs we use is within −40 to 85 °C. In our test cases, we use 100 and 110 °C, which are above the upper limit of temperature, but not so high to prevent the chip from functioning. The FPGA chips are heated up or cooled down in a commercial thermal chamber, which allows temperature fluctuation of ±0.3 °C. The core voltage is provided by a DC power supply and its nominal value is 1.2 V; the elevated voltage being used is within the 10% range and is much lower than the breakdown voltages. Figure 2.9 shows the measurement setup. The FPGA chip is placed in a zero insertion force (ZIF) socket board which connects with the interface board through

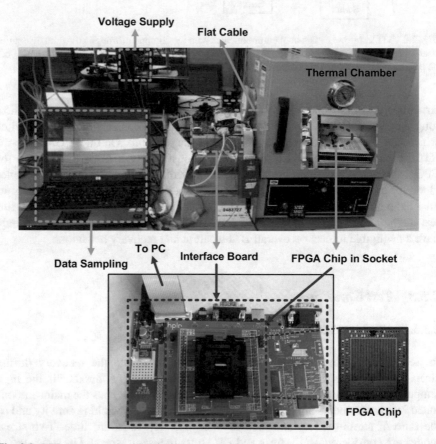

Fig. 2.9 BTI test setup: the FPGA chip is placed in a ZIF socket board, which connects with the interface board through a flat cable to separate the micro-controller from the high temperature environment, only the FPGA chip is in thermal chamber

a flat cable to ensure that only the FPGA chip is exposed to a high temperature environment.

2.5.3.3 Test Cases

All tests are carried out on a group of fresh FPGA chips. Several test cases in both stress (wearout) and recovery (including accelerated active self-healing) phases are considered and are denoted as follows (AS—accelerated stress, AR—accelerated recovery):

- **AS110AC24**: In this accelerated stress test case, the chip is under 110 °C environment for 24 h in AC stress mode. RO is always enabled to switch.
- **AS110DC24**: This is similar to the previous case, but in DC stress mode. RO is enabled only every 20 min for data recording. Data sampling overhead is less than 3 s.
- **AS100DC24**: 100 °C is applied and the chip is under accelerated DC stress mode for 24 h.
- **R20Z6**: In this case, chips are recovered for 6 h under 20 °C at 0 V.
- **AR20N6**: Negative voltage of -0.3 V is applied to the chip to activate the recovery at 20 °C.
- **AR110Z6**: In this case, only high temperature (110 °C) is applied, and the chip is powered off at 0 V for 6 h.
- **AR110N6**: Chips are recovered with both 110 °C and -0.3 V.

During recovery, a negative voltage and/or a high temperature of 110 °C are applied. During both stress and recovery periods, the test structure is enabled from the stress phase every 30 min for data sampling. The chip that is stressed and recovered under normal conditions ($T = 20$ °C, $V_{dds} = 1.2$ V, and $V_{ddr} = 0$ V) is used as the baseline for comparisons. All test cases are summarized in Table 2.1.

2.5.4 Modeling BTI Stress and Recovery for FPGA Test Structures

In Sect. 2.4.2, a gate-level analytic model that converts the device-level parametric shifts to circuit-level metric changes (delay change) was discussed. To analytically understand how BTI accelerated stress and recovery can affect the specific test

Table 2.1 Summary of test cases for accelerated stress and self-healing

Condition	Case index	Chip no.	T (°C)	V_{dd} (V)	Time (h)	Mode	Stress time / Recovery time
Stress (active)	AS110AC24	1	110	1.2	24	AC	–
	AS110DC24	2	110	1.2	24	DC	–
	AS110DC24	3	110	1.2	24	DC	–
	AS100DC24	4	100	1.2	24	DC	–
	AS110DC24	5	110	1.2	24	DC	–
	AS110DC48	5	110	1.2	48	DC	–
Recovery (sleep)	R20Z6	2	20	0	6	–	4
	AR20N6	3	20	−0.3	6	–	4
	AR110Z6	4	110	0	6	–	4
	AR110N6	5	110	−0.3	6	–	4
	AR110N12	5	110	−0.3	12	–	4

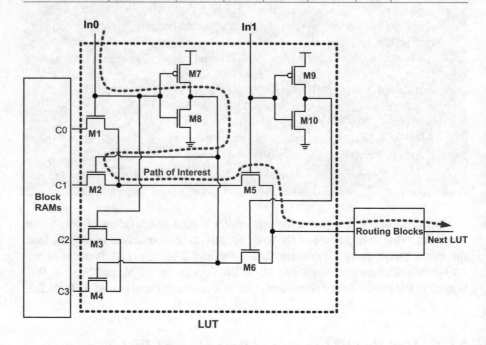

Fig. 2.10 Pass-transistor based LUT structure

structure on FPGA we chose, we apply the gate-level model to look-up-table (LUT) circuit; the details are as follows.

As shown in Fig. 2.6, the basic building blocks of the FPGA core are the LUTs, which are mapped as inverter logic in our BTI test structure (Fig. 2.7). Shown in Fig. 2.10 is a generic pass transistor (PT)-based 2-input LUT structure. Routing blocks include all the routing elements between LUT blocks. Four configuration bits ($C0$–$C3$) are stored in block RAM (BRAM), $In0$ and $In1$ are input signals. Let's consider an inverter mapped to a LUT: $In0$ is the input of the inverter, $C0$–$C3$ are 0101, and $In1$ is always 1. As shown in the figure, the path of interest (POI) is from the input of the LUT-based inverter to the output of the routing blocks. Assume the inverter is under DC stress, and $In0$ is always 1. $M1$, $M5$ are under stress and the threshold shift will affect the delay of POI. If $In0$ is always 0, only $M7$ is under stress. Based on this simple example, two hypotheses can be made:

- Not all the transistors on POI are under stress. In DC stress mode, once the inputs are given, the number of stressed and unstressed transistors is constant;
- Recovery can only have an impact on stressed transistors, but has no effect on "fresh" (never aged) transistors nor on transistors that have already recovered (close) to the "fresh" state.

Although the exact gate-level netlist of commercial FPGAs is not available, we believe that the two hypotheses above can be applied to any generic pass-transistor LUT configurations.

Assume that all stressed transistors on the POI are under the same stress condition (V_{gs} are the same), so we can approximately assume that ΔV_{th} of all stressed transistors are the same. The total delay change ΔT_d of POI becomes:

$$\Delta T_d = \sum_n^{LD} \Delta t_{dn} \sim \Delta t_d N_s \qquad (2.11)$$

where LD is the logic depth, N_s is the number of transistors that are under stress and $0 \leq N_s \leq LD$, and Δt_d is the delay change for a single gate from Eq. (2.6). Combine Eqs. (2.6) and (2.11), and assume $V_{dds} \gg V_{th}$, the total delay shift of a whole path after a stress period of t_{st} can be expressed as:

$$\Delta T_d(t_{st}) = Y exp\left(\frac{-E}{kT}\right) exp\left(\frac{BV_{dds}}{kTt_{ox}}\right)(A + log(1 + Ct_{st})) \qquad (2.12)$$

$$Y \sim K_{st}N_s t_{d0} \qquad (2.13)$$

If V_{dds} and T are constant over stress duration, Eq. (2.12) can be expressed as:

$$\Delta T_d(t_{st}) \sim \beta(A + log(1 + Ct_{st})) \qquad (2.14)$$

where β, A, and C are fitting parameters and can be extracted from measurement results. Similarly, during accelerated active recovery phase, we combine Eqs. (2.8), (2.9), and (2.11) and calculate the delay change of the POI after the recovery period t_{rec} as:

$$\Delta T_d(t_{st} + t_{rec}) = \frac{\phi_{rec}}{V_{dds}} \left(A + log(1 + Ct_{rec}) \right) + \Delta T_d(t_{st})$$

$$\times \left(1 - \frac{A + log(1 + Ct_{rec})}{A + log(1 + C(t_{st} + t_{rec}))} \right) \qquad (2.15)$$

Assume that the ratio of operation time to active sleep time of the system is α, the overall delay change in one cycle will be:

$$\Delta T_d(t_{total}) = \Phi_{acce} \left(A + log \left(1 + C \frac{t_{total}}{1+\alpha} \right) \right) + \Delta T_d \left(\frac{\alpha t_{total}}{1+\alpha} \right)$$

$$\times \left(1 - \frac{A + log \left(1 + C \frac{t_{total}}{1+\alpha} \right)}{A + log(1 + Ct_{total})} \right) \qquad (2.16)$$

$$\Phi_{acce} \sim K_{acce} exp \left(\frac{-E}{kT_{acce}} \right) exp \left(\frac{BV_{ddr}}{kT_{acce}t_{ox}} \right) \qquad (2.17)$$

The model evaluation and validation will be discussed together with the measurement results in the following sections.

2.6 Test Results for Accelerated BTI Wearout

This section presents the testing results during the *stress* period. It shows the performance degradation under various stress conditions.

2.6.1 AC Stress vs. DC Stress

AC stress and DC stress are conducted in the first and second case described in Sect. 2.5.3.3, Fig. 2.11 shows the measurement results at 110 °C. The AC stress with a 50% duty cycle can be treated as a symmetric stress vs. passive recovery case. In the first 3 h, RO frequency degradation of both cases is relatively fast and then becomes slower. AC stress has a symmetric stress and recovery process, during which stress phases are always followed by recovery phases due to dynamic activity of the circuit, and results in smaller frequency degradation, which is about half of that in the DC stress case. The results indicate that passive recovery is much slower compared to wearout since the chip cannot be fully recovered with symmetric AC stress. In other words, AC stress is an only partially self-healing process with a

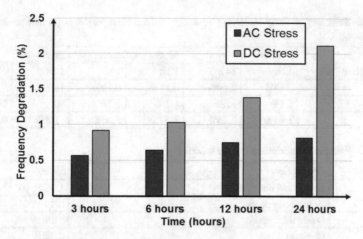

Fig. 2.11 AC and DC stress measurement results: AC stress case with a 50% duty cycle shows slower BTI wearout

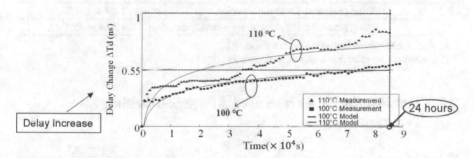

Fig. 2.12 Accelerated BTI wearout under 110 and 100 °C for 1 day: temperature has a significant impact on accelerating BTI Wearout

very slow recovery rate. To almost fully rejuvenate the chip, accelerated active self-healing techniques are required.

2.6.2   Effect of Temperature on BTI Wearout

Figure 2.12 shows measured delay change over time at 100 °C and 110 °C. As the model predicts, initially, frequency degrades fast and then slower. High temperature accelerates the degradation. Table 2.2 summarizes the delay change (%) for different temperature conditions. Table 2.3 shows the extracted parameters we used in the model that was discussed in Sect. 2.5.4.

Table 2.2 Summary of delay increase (%) under different temperatures (model prediction vs. measurement)

Temperature (°C)	Measurement		Model prediction		Error	
	12 h	24 h	12 h	24 h	12 h	24 h
20	0.13%	0.19%	0.11%	0.18%	15.3%	5.3%
100	1.1%	1.5%	1.1%	1.53%	0	2%
110	1.45%	2.16%	1.57%	1.96%	8.3%	9.3%

Table 2.3 Parameter descriptions for the model

Parameter	Description	Value
k	Boltzmann constant	1.38×10^{-23} J/K
E	Activation energy	0.49 eV
t_{ox}	Oxide thickness	1 nm
A^a	Constant	0.2801
B^a	Constant	3×10^{-29}
C^a	Constant	0.8614
K_{st}^b	Fitting parameter for stress equation	4.7×10^{-4}
K_{acce}^b	Fitting parameter for accelerated recovery equation	7.34×10^{-5}

[a] A, B, C are constants within the same technology
[b] K parameters are extracted based on measurement results

2.7 Test Results for Accelerated Active Self-healing Techniques

This section presents test results for the proposed BTI accelerated active recovery techniques. To make meaningful comparisons, we use recovered delay (delay decrease during recovery) as our metric for recovery, which can be calculated as $RD(t_2) = T_d(t_1) - T_d(t_2) = \Delta T_d(t_1) - \Delta T_d(t_2)$, where $t_2 > t_1$, and ΔT_d is the delay change with respect to the delay at time zero.

2.7.1 Active Recovery with Negative Voltage

On most of the current modern computing platforms, when the systems go to sleep, the supply voltage V_{dd} is usually gated to reduce leakage, but this only results in *passive* recovery. For *active recovery* we apply a negative voltage as the new supply voltage. The challenges for picking the value of this negative voltage are:

- Breakdown voltage limitation: the voltage must be at the level below the lateral PN-junction breakdown voltage;
- Implementation feasibility: implementation of negative voltage will introduce area and power overhead;

- GIDL: gate-induced drain leakage current (GIDL) may introduce unexpected leakage currents.

In our test cases, we picked a negative voltage of −0.3 V which has been validated to be still within the "safe" margin of the breakdown voltage and leakage. Figure 2.13 compares the recovered delay over 6 h (1/4 of the total stress time) when the temperature is set at 20 °C and 110 °C, respectively. Model predictions based on Sect. 2.4.2 are also included in the figure. The coefficients used in the model are all extracted from the measurements and are shown in Table 2.3. The test results in Fig. 2.13 show that stressed chips rejuvenate faster with a negative supply voltage for both temperatures. By applying a negative voltage the recovery is significantly accelerated even at room temperature.

Fig. 2.13 Negative voltage-enabled active recovery after being stressed for *24* h (net delay increase is ∼3.24 ns). *X*-axis is the recovery time. (**a**) At 20 °C, (**b**) at 110 °C

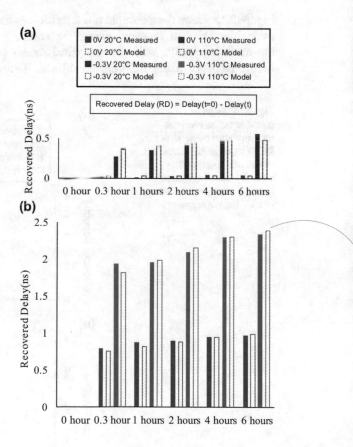

2.7.2 Accelerated Recovery with High Temperature

High temperature not only accelerates wearout, measurement results in this section show that it also accelerates recovery. Figure 2.14 presents the *recovered delay* vs. *temperature* test results under *passive recovery* (0 V) and *active recovery* (−0.3 V) conditions; in both cases, high temperature accelerates BTI recovery. The analytic model accurately predicts this behavior, as also shown in the figure.

2.7.3 Model Validation

Figure 2.15 shows the measured raw data (delay change ΔT_d) vs. model predictions over time for four recovery cases, the results indicate that test results match the modeling results well, the analytical model provides an accurate estimation of the recovery under different conditions. Table 2.4 summarizes the recovery

Fig. 2.14 High temperature-accelerated recovery after being stressed for *24* h (net delay increase is ∼3.24 ns). *X*-axis is the recovery time. (**a**) Under 0 V (*passive* recovery), (**b**) under −0.3 V (*active* recovery)

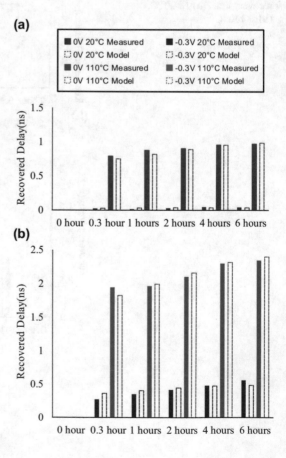

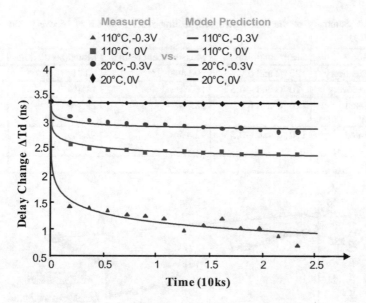

Fig. 2.15 Delay change (ΔT_d) over time during recovery: model predictions vs. measurement. Solid lines are model predictions, markers are measured data

percentage (recovered delay/net delay increase) under four test cases. Both model prediction and measurement results are included. It is worth mentioning that in the accelerated active recovery case when both high temperature and negative voltage are applied to the system, about 72.4% of the degradation can be recovered within only 1/4 of the stress time, and this also indicates that high temperature and negative voltage can "assist" each other during recovery and lead to the maximum recovered percentage (72.4% >> 16.7% + 28.7%)—a clear case of the combined effect exceeding the sum of the parts. The recovered percentage can be directly translated to the necessary design margin reduction. For example, with the accelerated and active recovery techniques, the design margin can be brought back to 27.6% of the original one within only a short period of recovery time. Figure 2.16 shows an example where the measured frequency over the whole period of wearout and accelerated self-healing behaviors is plotted, the recovery is under high temperature (110 °C), negative voltage (−0.3 V), and stress vs. recovery ratio of 4. In summary, accelerated active self-healing techniques for BTI are much more effective compared to the purely passive recovery condition, but we further observe that if we continue to apply accelerated active self-healing period after 12 h, BTI wearout can't actually be fully recovered and reaches a saturation even under the accelerated and active recovery conditions due to the existence of permanent components. In the following sections, we explore how to deal with this issue.

Table 2.4 Summary of the accelerated self-healing results for 6 h of recovery (%: recovered percentage)

Test case	Sleep condition	Measurement results	Model prediction	Error
Passive recovery	20 °C and 0 V	0.66%	0.71%	7.6%
Active recovery	20 °C and −0.3 V	16.7%	14.4%	13.8%
Accelerated recovery	110 °C and 0 V	28.7%	29.2%	1.7%
Accelerated active recovery	110 °C and −0.3 V	72.4%	72.7%	0.3%
Average				5.85%

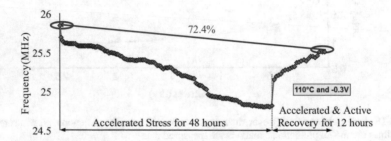

Fig. 2.16 One test case showing that accelerated self-healing techniques can recover about 72.4% of BTI wearout within only 1/4 of the stress time. Passive recovery data are not shown in the figure, but it is about less than 2% of recovery after 12 h

2.8 Reversible vs. Irreversible BTI Wearout

Our experiments show that accelerated self-healing techniques are able to rejuvenate the "aged" chip from BTI effectively. While based on the physical trapping and detrapping mechanisms of BTI discussed in Sect. 2.2, and also validated by our experiments, there have been still lingering irreversible components (e.g., shown in Fig. 2.3 as V_{th} *net still manage to increase*) that are hardly recovered within a reasonable time and accumulate cycle by cycle, thus hurting the performance and increasing the necessary design guardband. In this section, we look into the permanent component of BTI more in detail and explore the frequency dependency of wearout and recovery. A similar experimental setup and test methodology used in Sect. 2.5 are also employed in this part of the work.

2.8.1   Fast Traps vs. Slow Traps: A Physics Perspective

The BTI mechanisms [2, 8] suggest that the charge carriers need to gain sufficient kinetic energy to overcome a potential barrier necessary to break an interface state to be captured in the traps during the trapping process. Here we define *fast traps* as those traps that have a higher probability of trapping the charge carriers. These traps have a relatively lower trap energy barrier and are easier to be filled in a shorter time. On the contrary, *slow traps* are those that have a higher trap energy barrier that is difficult for charge carriers to overcome. The principle of physics for BTI detrapping (recovery process) is that the trapped charge carriers (e.g., electrons for NMOS or holes for PMOS) have a certain probability to escape, with the probability being higher if their energy is higher and the trap energy barrier lower, and vice versa. Based on the statistical mechanics theory, the distribution of kinetic energies is thus proportional to the product of density of state and the Boltzmann distribution [46]. The three-dimensional density of state is proportional to $\sqrt{E}$; therefore, the energy distribution of electron is given by:

$$f_E(E) = A \times \left(\frac{1}{kT}\right)^{3/2} \times \sqrt{E} \times exp\left(-\frac{E}{kT}\right) \qquad (2.18)$$

where E is the energy of the electrons, k is Boltzmann constant, T is temperature in Kelvin, and A is a normalization factor. Since the detrapping rate is proportional to the number of electrons at the energy of consideration, the detrapping rate is proportional to $f_E(E)$. The energy distribution of electrons at room temperature (300 K) is plotted in Fig. 2.17, which shows that a majority of the electrons are at low energy in meV range, whereas the center energy of even the lowest energy of the trap is in the order of several kT ($\sim$0.026 eV) [47, 48]. This means that only a small fraction of electrons at the tail of the distribution could participate in the detrapping process.

Shown in Fig. 2.18 is an illustration of the trapping and detrapping process for two types of the traps. Since fast traps have lower trap energy barrier, it is easier for charge carriers escape from them; this leads to *fast recovery*, or the *reversible* component of wearout. For the slow traps, the charge carriers need to overcome a higher trap energy barrier. As a result, it is very slow or even impossible for them to escape within a reasonable time, so these traps will cause the *irreversible* component

Fig. 2.17 Energy distribution of electrons at room temperature

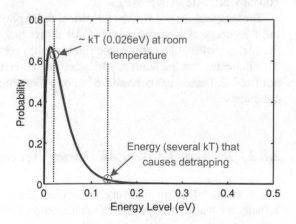

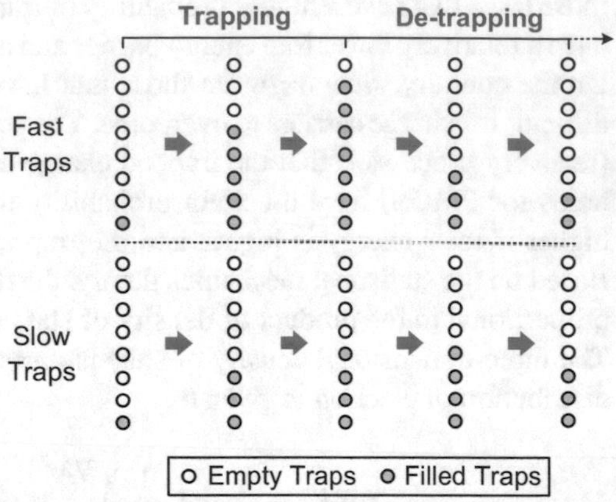

Fig. 2.18 Illustration of fast
traps vs. slow traps

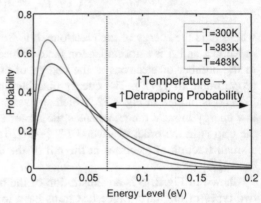

Fig. 2.19 Energy distribution
of electrons at room
temperature

of wearout. To give a first-order estimation, if the trap energy is 100 meV higher, the
probability distribution goes down by a factor of $exp(100\,\text{meV}/20\,\text{emV}) \approx 50$. This
indicates that the time it takes to detrap goes up by about a factor of 50 for every
100 meV increase in trap energy.

The temperature and voltage (electrical field) play an important role in determining the energy of electrons. Figure 2.19 shows that by increasing the temperature,
the energy distribution shifts to the right, so the probability of detrapping increases.
This indicates that the boundary between the reversible and irreversible wearout is
not fixed and can actually be shifted in one direction or the other depending on the
conditions.

2.8.2   Irreversible Wearout During Accelerated Self-healing

The accelerated active recovery techniques discussed in Sect. 2.7.3 are able to
actually recover a small portion of what would otherwise be considered *irreversible*

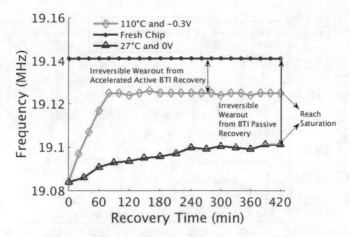

Fig. 2.20 Irreversible part under two recovery conditions—passive recover vs. accelerated active recovery (accelerated self-healing). BTI doesn't recover completely even under the accelerated self-healing conditions

wearout. This is demonstrated by our measured experimental results that are shown in Fig. 2.20. The frequency under passive recovery condition (27 °C and $V_{gs} = 0$ V) is normalized to the accelerated self-healing condition. It is clear that in both cases, recovery saturates to some values below the fresh state frequency, and the irreversible part of BTI wearout after passive recovery is much larger than for the accelerated and active recovery case.

The irreversible components that are left unchecked during recovery keep building up throughout the rest of the system lifetime. Figure 2.21 demonstrates exactly this; it shows a test case where several cycles of stress and recovery are scheduled. In each cycle, a 6-h accelerated stress is followed by a 6-h accelerated recovery (6 h vs. 6 h). IRn refers to the amount of irreversible wearout for the nth cycle. The figure shows that the recovery under accelerated conditions saturates in each cycle, and the irreversible BTI component (IR) increases for the first few cycles and settles afterwards. A possible explanation for this behavior is that in the later cycles, some of the irreversible wearout from previous cycles starts to recover, and the accelerated recovery and stress can fully compensate each other. But the wearout will still not be fully recovered to the fresh state even with intense accelerated active recovery techniques applied in each cycle.

2.8.3 Sequentiality of Reversible and Irreversible Wearout

Reversible wearout and irreversible wearout are mostly determined by *fast* traps and *slow* traps, respectively. This suggests that different stress/recovery sequences will have different wearout "profiles" (ratio of reversible to irreversible) due to the

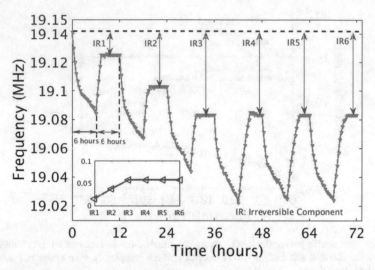

Fig. 2.21 Accumulation of the irreversible BTI wearout cycle by cycle. IRn refers to the accumulated irreversible component after nth cycle

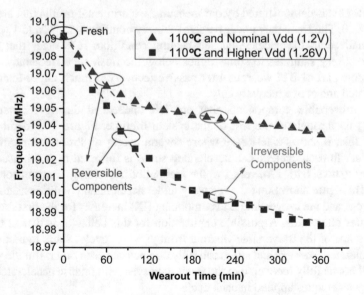

Fig. 2.22 BTI-induced frequency degradation under two accelerated stress conditions. In both cases, reversible wearout kicks in firstly, then it levels off and irreversible wearout takes over

different trapping rates of the two. To further investigate this, a set of stress tests is conducted. Shown in Fig. 2.22 is the frequency degradation under two accelerated stress conditions with different stress voltages. It demonstrates that both test cases follow similar wearout patterns. Firstly, reversible wearout kicks in, then the effect of reversible wearout levels off and irreversible wearout takes over—in the time

Fig. 2.23 Sequentiality of
reversible and irreversible
wearout

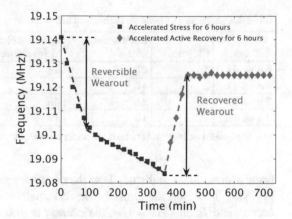

domain this is roughly seen as a steep slope followed by a shallow slope during
wearout.

Figure 2.23 shows a test case where a 6-h accelerated and active recovery
(110 °C, −0.3 V) duration follows a 6-h of accelerated stress. The recovery process
speeds up as expected, and even some parts that would otherwise been considered
irreversible start to recover. In the time domain this can be roughly seen as a steep
slope followed by a *saturation* (zero slope) once all the reversible part and part
of irreversible part were recovered. Also, it is worth to mention that the *recovered
wearout* is larger than the *reversible wearout* as shown in the figure, and this further
demonstrates that the *accelerated and active recovery* techniques are able to recover
some of the irreversible parts. But solutions that are able to fully fix or avoid this
component are still necessary and highly preferred; they are explored in the next
section.

2.9   Frequency Dependency of BTI Wearout and Recovery

2.9.1   Sleep with Accelerated Rejuvenation When Getting Tired

The whole process of wearout and accelerated active recovery can be compared
directly to the biological world. Humans, for example, are active during daytime,
with the body conducting activities and experiencing fatigue. During night-time
sleep, the body goes through several active processes that are essential for the
recovery of its full capabilities for the next day. If some organs experience heavy
fatigue without in-time rest, part of the fatigue will be translated into potential
permanent harm to the body, and will be hardly recovered. This is well known for
athletes who need scheduled recovery periods after extensive workouts, with their
athletic performance actually getting better after the rest periods. These biological
fatigue and recovery schedules are not unlike those shown in Fig. 2.21. We thus

Table 2.5 Summary of periodic accelerated active recovery test cases[a]

Test case	Chip no.	Cycle stress time	Cycle accelerated active recovery time	# of cycles	Total time
6 h vs. 6 h	1	6 h	6 h	6	3 days
4 h vs. 4 h	2	4 h	4 h	9	3 days
2 h vs. 2 h	3	2 h	2 h	18	3 days
1 h vs. 1 h	4	1 h	1 h	32	3 days

[a]All accelerated active recovery tests are under the −0.3 V and 110 °C condition

borrow these ideas and study how they can be applied to electronic systems. The key idea is to stop the stress before irreversible effects get a chance to start to accumulate. *In an ideal setting, the strategy is to keep the circuit active only during the reversible phase of BTI wearout until the irreversible part kicks in; thus, the irreversible wearout becomes almost unobservable after active recovery even under accelerated stress cases. In this way, the system follows an internal circadian rhythm to recover for its full computing capability.* To validate this idea, a set of tests with different "circadian rhythms"1 is conducted, and is summarized in Table 2.5. All tests start with the FPGAs in the fresh state. The total test time is 3 days for all test cases.

To make a fair comparison of recovery percentage among different chips and also under different operating conditions, some normalization is necessary. From Eqs. (2.4) and (2.5), we can estimate the change in gate delay Δt_d when V_{th} is subject to change as:

$$\Delta t_d \sim \Gamma \cdot \Delta V_{th} \cdot t_{g0} \tag{2.19}$$

where t_{g0} is the time zero delay of the gate with no V_{th} shift, Γ is a constant. Assume that chip 1 has an initial frequency (fresh status) of $f_{1T_A}(0)$ at temperature T_A, and chip 2 has an initial frequency of $f_{2T_B}(0)$ at temperature T_B. If two chips undergo the same threshold voltage change $\Delta V_{th}(t)$ after being stressed or recovered for time length of t, and since the temperature induced threshold voltage shift doesn't change with time, so the resulted frequency of chip 1 at temperature T_A becomes:

$$f_{1T_A}(t) \sim \frac{1}{t_{1T_A}(0) + \Gamma \cdot \Delta V_{th}(t) \cdot t_{1T_A}(0)} \tag{2.20}$$

Similarly, the frequency of chip 2 at temperature T_B is:

$$f_{2T_B}(t) \sim \frac{1}{t_{2T_B}(0) + \Gamma \cdot \Delta V_{th}(t) \cdot t_{2T_B}(0)} \tag{2.21}$$

The process of normalizing chip 2 data to chip 1 data is done through

$$\frac{f_{2T_B}(t)}{f_{2T_B}(0)} \cdot f_{1T_A}(0) = \frac{t_{2T_B}(0)}{t_{2T_B}(0) + \Gamma \cdot \Delta V_{th}(t) \cdot t_{2T_B}(0)} \cdot \frac{1}{t_{1T_A}(0)}$$

$$= \frac{1}{t_{1T_A}(0) + \Gamma \cdot \Delta V_{th}(t) \cdot t_{1T_A}(0)} = f_{1T_A}(t) \quad (2.22)$$

The above derivation proves that although the same amount of BTI-induced threshold voltage $\Delta V_{th}(t)$ could lead to different frequency degradation for different chips under different temperatures, the normalization process will reflect the actual threshold voltage shift across multiple chips properly. Thus, in the following measurement results section, we are able to normalize the frequency of all four test cases shown in Table 2.5 to be able to make fair comparisons.

2.9.2 Measurement Results

Shown in Fig. 2.24 are measurement results under four circadian rhythms. Accelerated active recovery with both high temperature and negative voltage is applied in each recovery interval, it shows that for all test cases except the 1 h vs. 1 h case, the accelerated active recovery has a period of saturation which indicates the irreversible parts of the wearout, and the irreversible parts accumulate in several of the initial cycles and saturate in the following cycles. For the 1 h vs. 1 h case, alternating phases of stress and accelerated recovery can almost completely compensate for each other, and after each accelerated active recovery phase the chip can indeed start fresh. *The irreversible part of wearout is thus close to be fully recovered.* Figure 2.25 presents the accumulated irreversible wearout for the first six cycles under the four test conditions. It shows that the earlier the accelerated active recovery conditions are applied, the slower the irreversible wearout will accumulate, which leads to less permanent component. There is an optimal balance of stress and accelerate recovery (e.g., 1 h vs. 1 h in this accelerated case) which can result in almost no irreversible wearout.

A 1 h vs. 1 h may seem too short for practical applications, but remember that these tests are under accelerated stress conditions; the equivalent normal operation durations are much longer. For example, assume that the amount of frequency degradation under normal condition is the same as for the accelerated case of 1 h, the equivalent duration is about *31 h* under nominal voltage and room temperature based on the model proposed in Sect. 2.4.2. As shown in Fig. 2.26, the identical optimal condition could be that the chip is active under normal operating conditions for (at most) *31* h which is quite attractive for practical applications. The accumulated BTI wearout could be still fully mitigated by a following *1* h (or longer) accelerated active recovery duration. Details about how to apply these behaviors in a real system will be discussed in Chap. 5.

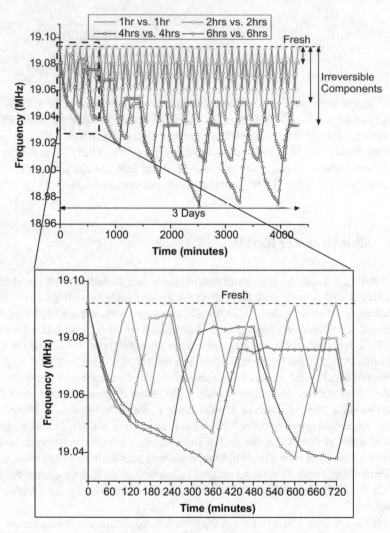

Fig. 2.24 Measurement results under four different "Circadian Rhythms." The *1 h vs. 1 h* case leads to that the irreversible BTI component is almost completely avoided and eliminated

2.9.3 Reduction of Necessary Design Margin

The explored unique behaviors of BTI wearout and (accelerated and active) recovery can lead to significant reduction of the necessary static design margins that are added in the early design phase. As discussed in Chap. 1, for the worst-case design solution, to meet the timing requirement throughout the whole lifetime, guardbands need to be added (e.g., by oversizing transistors). Without the scheduled accelerated active recovery under an optimal circadian rhythm, the margin needs to cover both

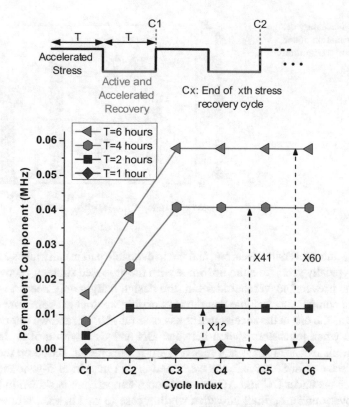

Fig. 2.25 Measured irreversible BTI components (permanent components) that are accumulated during the first six cycles for four test cases under four different scheduled accelerated stress and recovery circadian rhythms

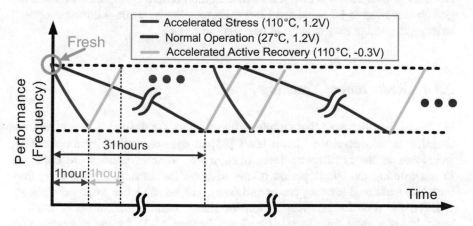

Fig. 2.26 An identical regular-operation use case (31 h vs. 1 h) to the 1 h vs. 1 h accelerated stress case for FULL recovery

Fig. 2.27 Necessary design
margin estimation under
different stress conditions

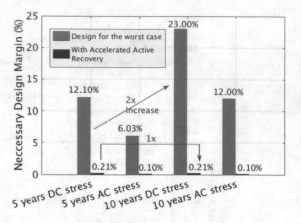

reversible and irreversible wearout, and the irreversible part usually takes a long time
period (typically years) to build up. Since with the proposed strategy, recovery starts
before the irreversible wearout kicks in, the design margin only needs to cover the
reversible component. Assume that the irreversible wearout at room temperature is
the same as the one at the accelerated stress case (at 110 °C), and we refer AC stress
as a case when transistors switch between ON and OFF with a 50% duty cycle,
which means balanced stress and recovery during operation. Based on the analytic
model discussed in Sect. 2.4.2, the estimated design margin of 5-year and 10-year
lifetime spans under DC and AC stress at room temperature is shown in Fig. 2.27.
The recovery under optimal circadian rhythm case (e.g., 1 h. vs. 1 h case) gives a
design margin reduction of more than 60× for all cases. In the AC stress case for
a 10-year lifetime constraint, the design margin reduction also scales up and can
reach more than 100×. It also shows that as the lifetime constraint increases, the
guardbands need to be expanded (2×) correspondingly, while with the proposed
strategy, the design margin stays almost the same (1×).

2.9.4   Reduction of Tracking Power

In Chap. 1, we discussed that an alternative solution for dealing with wearout are
adaptive techniques at the circuit level [23] or dynamic reliability management
techniques at the architecture level [49], where wearout sensors are deployed
to track during the whole period of the lifetime. But sensors are not free, they
introduce additional tracking power and area overhead. With the proposed recovery
strategy, the time for recovery is known ahead, wearout sensors only need to
track during a short time (e.g., 31 h shown in Sect. 2.9.1) for the reversible part
of wearout. Numerically, the difference for the tracking power is estimated as

between $O(ln(31\,h))$ and $O(ln(10\,\text{years}))$ for a 10-year lifetime constraint, or about $ln(2826) \sim 8\times$ reduction.

2.9.5 Average Performance Improvement

With the periodic accelerated and active recovery, the circuit is guaranteed to run faster than the case when no recovery is applied. Here, we define that the average performance refers to the average of all performance values across time during operation (wearout). Figure 2.28 shows the average performance improvement (IMP) calculated based on the measurement results for 1 day and 2 days for the same chip. Under the 1 h vs. 1 h case, when irreversible wearout is almost fully recovered, it gives the best average performance, which is close to the fresh status. As operation time increases (e.g., from 1 day to 2 days), the average performance will keep almost the same for 1 h vs. 1 h case, while for other test cases, especially the case when no recovery strategy is applied, the average performance decreases dramatically, and this concludes that the average performance improvement achieved by the proposed strategy will scale up with time (e.g., $1.6\times$ from 1 day to 2 days as shown in the figure).

Figure 2.29 presents the predicted average performance improvement over the no-recovery solution under nominal operation conditions (room temperature, nominal voltage) predicted by the model. As the lifetime constraint increases from 5 years to 10 years, the average performance for the accelerated active recovery solution will keep almost the same, compared to the no-recovery case when the average performance reduces dramatically. In other words, the average performance improvement enabled by recovery will increase significantly as the lifetime

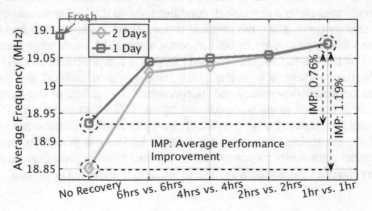

Fig. 2.28 Average performance improvement (IMP) for 1 day and 2 days based on measurement results

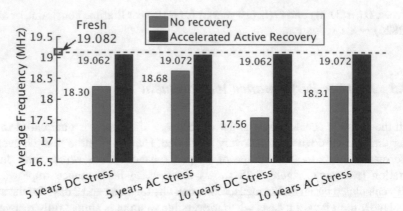

Fig. 2.29 Average performance under different stress conditions

requirement increases. To give an example, for a 10-year lifetime span, the improvement can be as large as ~9%.

2.9.6 Frequency Dependency Behavior of BTI Wearout

It has been shown in many works [50, 51] that the BTI-induced V_{th} shift due to AC stress is independent of frequency in the range of few Hz to GHz, and most of the BTI models are also developed based on this theory, while our experimental results clearly demonstrate that the hypothesis of the independence of BTI on frequency doesn't hold for the whole frequency spectrum, especially in the "slow" frequency range. Figure 2.30 shows the measured permanent component vs. frequency under four accelerated stress and recovery conditions. As in the higher frequency range (close to 0 on the X-axis), the permanent component is almost zero and doesn't increase with frequency. When the frequency reaches the lower range (moving to the right on the X-axis), the accumulated permanent component increases inversely with frequency. There is a turning point (range) which "defines" the boundary between frequency dependency and independence. To the best of our knowledge, this book is the first to study experimentally the BTI effect under very slow AC stress with the frequency range of 1/h and also under accelerated stress and recovery conditions, the explored frequency dependent behaviors can potentially lead to the fact that the permanent component can be minimal through proactively scheduled recovery intervals. Implementation details are further discussed in Chap. 5.

Fig. 2.30 Frequency
dependency of irreversible
component of BTI wearout

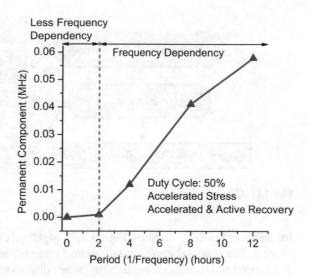

2.10 Conclusions

As BTI becomes one of the dominant reliability challenges for present and future
digital circuits, most of the previous BTI mitigation techniques focus on reducing
BTI-induced degradation during operation (under stress) or utilizing the passive
recovery behavior of BTI. However BTI is not fundamentally repaired in those
cases, and this can cause permanent failures. In this chapter, we first presented
a series of accelerated self-healing techniques which are able to "reverse" the
direction of wearout, thus accelerating and activating the BTI recovery. Based on
the actual hardware measurement results with commercial FPGAs, these accelerated
and active self-healing techniques can lead to significant recovery rate improvements (e.g., we demonstrated a case where 72.4% of the wearout is recovered
within only 1/4 of the stress time for BTI). Still, even under the accelerated
active recovery conditions, there are still parts of BTI wearout that are irreversible;
thus, we further explored the boundary between reversible and irreversible wearout
physically and experimentally. The main findings are: first, we show that the
boundary between the reversible and irreversible parts of wearout is not fixed, with
the irreversible part becoming at least partially reversible under the right conditions
of accelerated active recovery; second, we show that there are certain stress/recovery
schedules that can (almost) completely eliminate irreversible wearout, thus allowing
significant reductions in necessary design margins (>60×) and improvement in
average performance (∼9% with a 10-year lifetime). The discovered BTI frequency
dependence (at slow frequencies) is able to help the community understand BTI
behaviors more deeply, these unique BTI recovery properties introduce a new knob

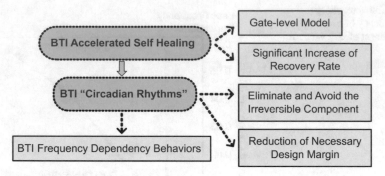

Fig. 2.31 Chapter 2 highlights

for designing reliable systems. Figure 2.31 highlights the key contributions of this chapter. The potential implementations and trade-off analysis by making use of the accelerated active self-healing techniques are discussed in the following chapters.

References

1. S Mahapatra, V Huard, A Kerber, V Reddy, S Kalpat, and A Haggag. Universality of nbti-from devices to circuits and products. In *Reliability Physics Symposium, 2014 IEEE International*, pages 3B–1. IEEE, 2014.
2. Souvik Mahapatra. *Fundamentals of Bias Temperature Instability in MOS Transistors.* Springer, 2016.
3. Wenping Wang, Shengqi Yang, Sarvesh Bhardwaj, Rakesh Vattikonda, Sarma Vrudhula, Frank Liu, and Yu Cao. The impact of nbti on the performance of combinational and sequential circuits. In *Proceedings of the 44th annual Design Automation Conference*, pages 364–369. ACM, 2007.
4. James H Stathis, Souvik Mahapatra, and Tibor Grasser. Controversial issues in negative bias temperature instability. *Microelectronics Reliability*, 81:244–251, 2018.
5. Souvik Mahapatra and Narendra Parihar. A review of nbti mechanisms and models. *Microelectronics Reliability*, 81:127–135, 2018.
6. Rakesh Vattikonda, Wenping Wang, and Yu Cao. Modeling and minimization of pmos nbti effect for robust nanometer design. In *Proceedings of the 43rd annual Design Automation Conference*, pages 1047–1052. ACM, 2006.
7. Jyothi Bhaskarr Velamala, Ketul B Sutaria, Hirofumi Shimizu, Hiromitsu Awano, Takashi Sato, Gilson Wirth, and Yu Cao. Compact modeling of statistical bti under trapping/detrapping. *IEEE Transactions on Electron Devices*, 60(11):3645–3654, 2013.
8. Yu Cao, Jyothi Velamala, Ketul Sutaria, Mike Shuo-Wei Chen, Jonathan Ahlbin, Ivan Sanchez Esqueda, Michael Bajura, and Michael Fritze. Cross-layer modeling and simulation of circuit reliability. *Computer-Aided Design of Integrated Circuits and Systems, IEEE Transactions on*, 33(1):8–23, 2014.
9. V Huard, M Denais, and C Parthasarathy. Nbti degradation: From physical mechanisms to modelling. *Microelectronics Reliability*, 46(1):1–23, 2006.
10. Subrat Mishra, Hiu Yung Wong, Ravi Tiwari, Ankush Chaudhary, Narendra Parihar, Rakesh Rao, Steve Motzny, Victor Moroz, and Souvik Mahapatra. Predictive tcad for nbti stress-recovery in various device architectures and channel materials. In *Reliability Physics Symposium (IRPS), 2017 IEEE International*, pages 6A–3. IEEE, 2017.

11. A Benabdelmoumene, B Djezzar, A Chenouf, H Tahi, B Zatout, and M Kechouane. On the turn-around phenomenon in n-mos transistors under nbti conditions. *Solid-State Electronics*, 121:34–40, 2016.
12. S Pae, M Agostinelli, M Brazier, R Chau, G Dewey, T Ghani, M Hattendorf, J Hicks, J Kavalieros, K Kuhn, et al. Bti reliability of 45 nm high-k+ metal-gate process technology. In *Reliability Physics Symposium, 2008. IRPS 2008. IEEE International*, pages 352–357. IEEE, 2008.
13. S Zafar, YH Kim, V Narayanan, C Cabral Jr, V Paruchuri, B Doris, J Stathis, A Callegari, and M Chudzik. A comparative study of nbti and pbti (charge trapping) in sio2/hfo2 stacks with fusi, tin, re gates. In *VLSI Technology, 2006. Digest of Technical Papers. 2006 Symposium on*, pages 23–25. IEEE, 2006.
14. Tibor Grasser, B Kaczer, W Goes, Th Aichinger, Ph Hehenberger, and M Nelhiebel. A two-stage model for negative bias temperature instability. In *Reliability Physics Symposium, 2009 IEEE International*, pages 33–44. IEEE, 2009.
15. Narendra Parihar, Uma Sharma, Subhadeep Mukhopadhyay, Nilesh Goel, Ankush Chaudhary, Rakesh Rao, and Souvik Mahapatra. Resolution of disputes concerning the physical mechanism and DC/AC stress/recovery modeling of Negative Bias Temperature Instability (NBTI) in p-MOSFETs. In *Reliability Physics Symposium (IRPS), 2017 IEEE International*, pages XT–1. IEEE, 2017.
16. Jyothi Bhaskarr Velamala, Ketul Sutaria, Takashi Sato, and Yu Cao. Physics matters: statistical aging prediction under trapping/detrapping. In *Proceedings of the 49th Annual Design Automation Conference*, pages 139–144. ACM, 2012.
17. Jaume Abella, Xavier Vera, and Antonio Gonzalez. Penelope: The nbti-aware processor. In *Microarchitecture, 2007. MICRO 2007. 40th Annual IEEE/ACM International Symposium on*, pages 85–96. IEEE, 2007.
18. Xiaoming Chen, Yu Wang, Yu Cao, Yuchun Ma, and Huazhong Yang. Variation-aware supply voltage assignment for simultaneous power and aging optimization. *Very Large Scale Integration (VLSI) Systems, IEEE Transactions on*, 20(11):2143–2147, 2012.
19. Saket Gupta and Sachin S Sapatnekar. Gnomo: Greater-than-nominal v dd operation for bti mitigation. In *Design Automation Conference (ASP-DAC), 2012 17th Asia and South Pacific*, pages 271–276. IEEE, 2012.
20. Saket Gupta and Sachin S Sapatnekar. Employing circadian rhythms to enhance power and reliability. *ACM Transactions on Design Automation of Electronic Systems (TODAES)*, 18(3):38, 2013.
21. Mohammad Saber Golanbari, Nour Sayed, Mojtaba Ebrahimi, Mohammad Hadi Moshrefpour Esfahany, Saman Kiamehr, and Mehdi B Tahoori. Aging-aware coding scheme for memory arrays. In *Test Symposium (ETS), 2017 22nd IEEE*, pages 1–6. IEEE, 2017.
22. Abhishek Tiwari and Josep Torrellas. Facelift: Hiding and slowing down aging in multicores. In *Microarchitecture, 2008. MICRO-41. 2008 41st IEEE/ACM International Symposium on*, pages 129–140. IEEE, 2008.
23. Nimay Shah, Rupak Samanta, Ming Zhang, Jiang Hu, and Duncan Walker. Built-in proactive tuning system for circuit aging resilience. In *Defect and Fault Tolerance of VLSI Systems, 2008. DFTVS'08. IEEE International Symposium on*, pages 96–104. IEEE, 2008.
24. Taniya Siddiqua and Sudhanva Gurumurthi. Nbti-aware dynamic instruction scheduling. In *Proceedings of the 5th Workshop on Silicon Errors in Logic-System Effects*. Citeseer, 2009.
25. Lin Li, Youtao Zhang, Jun Yang, and Jianhua Zhao. Proactive nbti mitigation for busy functional units in out-of-order microprocessors. In *Proceedings of the Conference on Design, Automation and Test in Europe*, pages 411–416. European Design and Automation Association, 2010.
26. Dean Michael Ancajas, Koushik Chakraborty, and Sanghamitra Roy. Proactive aging management in heterogeneous nocs through a criticality-driven routing approach. In *Proceedings of the Conference on Design, Automation and Test in Europe*, pages 1032–1037. EDA Consortium, 2013.

27. Hans Reisinger, Oliver Blank, Wolfgang Heinrigs, Wolfgang Gustin, and Christian Schlünder. A comparison of very fast to very slow components in degradation and recovery due to nbti and bulk hole trapping to existing physical models. *Device and Materials Reliability, IEEE Transactions on*, 7(1):119–129, 2007.
28. Pradip Bose, Jeonghee Shin, and Victor Zyuban. Method for extending lifetime reliability of digital logic devices through removal of aging mechanisms, February 10 2009. US Patent 7,489,161.
29. Pradip Bose, Jeonghee Shin, and Victor Zyuban. Method for extending lifetime reliability of digital logic devices through reversal of aging mechanisms, February 3 2009. US Patent 7,486,107.
30. Jeonghee Shin, Victor Zyuban, Pradip Bose, and Timothy M Pinkston. A proactive wearout recovery approach for exploiting microarchitectural redundancy to extend cache sram lifetime. In *ACM SIGARCH Computer Architecture News*, volume 36, pages 353–362. IEEE Computer Society, 2008.
31. Taniya Siddiqua and Sudhanva Gurumurthi. Recovery boosting: A technique to enhance nbti recovery in sram arrays. In *VLSI (ISVLSI), 2010 IEEE Computer Society Annual Symposium on*, pages 393–398. IEEE, 2010.
32. Aditya Bansal and Jae-Joon Kim. Power napping technique for accelerated negative bias temperature instability (nbti) and/or positive bias temperature instability (pbti) recovery, July 21 2015. US Patent 9086865.
33. Thomas Aichinger, Michael Nelhiebel, and Tibor Grasser. On the temperature dependence of nbti recovery. *Microelectronics Reliability*, 48(8):1178–1184, 2008.
34. Anastasios A Katsetos. Negative bias temperature instability (nbti) recovery with bake. *Microelectronics Reliability*, 48(10):1655–1659, 2008.
35. Gregor Pobegen, Thomas Aichinger, Michael Nelhiebel, and Tibor Grasser. Understanding temperature acceleration for nbti. In *Proc. Intl. Electron Devices Meeting (IEDM)*, pages 27–3, 2011.
36. Tibor Grasser, Th Aichinger, Gregor Pobegen, Hans Reisinger, P-J Wagner, Jacopo Franco, M Nelhiebel, and Ben Kaczer. The 'permanent' component of nbti: composition and annealing. In *Reliability Physics Symposium (IRPS), 2011 IEEE International*, pages 6A–2. IEEE, 2011.
37. Boualem Djezzar, Hakim Tahi, Abdelmadjid Benabdelmoumene, Amel Chenouf, Mohamed Goudjil, and Youcef Kribes. On the permanent component profiling of the negative bias temperature instability in p-mosfet devices. *Solid-State Electronics*, 106:54–62, 2015.
38. Jan M Rabaey, Anantha P Chandrakasan, and Borivoje Nikolic. *Digital integrated circuits*, volume 2. Prentice hall Englewood Cliffs, 2002.
39. KK Ramakrishnan, Smitha Suresh, Narayanan Vijaykrishnan, Mary Jane Irwin, and Vijay Degalahal. Impact of nbti on fpgas. In *VLSI Design, 2007. Held jointly with 6th International Conference on Embedded Systems., 20th International Conference on*, pages 717–722. IEEE, 2007.
40. Yasuo Sato, Masafumi Monden, Yousuke Miyake, and Seiji Kajihara. Reduction of nbti-induced degradation on ring oscillators in fpga. In *Dependable Computing (PRDC), 2014 IEEE 20th Pacific Rim International Symposium on*, pages 59–67. IEEE, 2014.
41. Giray Kömürcü, Ali Emre Pusane, and Günhan Dündar. Effects of aging and compensation mechanisms in ordering based ro-pufs. *Integration, the VLSI Journal*, 52:71–76, 2016.
42. Lattice Semiconductor iCE40 HX-Series Ultra Low-Power mobile FPGA Family Datasheet:. http://www.latticesemi.com/Products/FPGAandCPLD/iCE40.aspx.
43. M Naouss and F Marc. Design and implementation of a low cost test bench to assess the reliability of fpga. *Microelectronics Reliability*, 55(9):1341–1345, 2015.
44. Siva Velusamy, Wei Huang, John Lach, Mircea Stan, and Kevin Skadron. Monitoring temperature in fpga based socs. In *Computer Design: VLSI in Computers and Processors, 2005. ICCD 2005. Proceedings. 2005 IEEE International Conference on*, pages 634–637. IEEE, 2005.
45. Atmel SAM7SE microcontroller Education Kit. http://www.atmel.com/tools/sam7se-ek.aspx.

46. Hasse Fredriksson and Ulla Akerlind. *Physics of functional materials*. John Wiley & Sons, 2008.
47. Vincent Huard, Florian Cacho, Xavier Federspiel, and Pascal Mora. Hot-carrier injection degradation in advanced cmos nodes: a bottom-up approach to circuit and system reliability. In *Hot Carrier Degradation in Semiconductor Devices*, pages 401–444. Springer, 2015.
48. Ricardo Reis, Yu Cao, and Gilson Wirth. *Circuit design for reliability*. Springer, 2015.
49. Jayanth Srinivasan, Sarita V Adve, Pradip Bose, and Jude A Rivers. The case for lifetime reliability-aware microprocessors. In *ACM SIGARCH Computer Architecture News*, volume 32, page 276. IEEE Computer Society, 2004.
50. Raoul Fernandez, Ben Kaczer, Axel Nackaerts, Steven Demuynck, R Rodriguez, Montserat Nafria, and Guido Groeseneken. Ac nbti studied in the 1 hz–2 ghz range on dedicated on-chip cmos circuits. In *Electron Devices Meeting, 2006. IEDM'06. International*, pages 1–4. IEEE, 2006.
51. GCKY Chen, KY Chuah, MF Li, Daniel SH Chan, CH Ang, JZ Zheng, Y Jin, and DL Kwong. Dynamic nbti of pmos transistors and its impact on device lifetime. In *Reliability Physics Symposium Proceedings, 2003. 41st Annual. 2003 IEEE International*, pages 196–202. IEEE, 2003.

Chapter 3
Accelerating and Activating Recovery Against EM Wearout

3.1 Overview

The downscaling of CMOS technologies into the nano-regime not only exacerbates transistor wearout issues such as BTI, it also worsens the interconnect (on-chip metal wire) electromigration (EM). Specifically, EM has become a significant concern in power delivery networks (PDN), which largely experience unidirectional current flow [1, 2]. EM-induced failure is projected to get even worse due to the increasing current densities from shrinking interconnect geometries in the sub-10 nm regime [3]. EM occurs due to the gradual displacement of metal atoms in wires as current passes through. When the current density is high enough, it can cause the drift of metal ions in the direction of the electron flow. While BTI degrades chip performance by slowing down transistor switching speed, EM increases wire resistance, which causes voltage drop, thus also resulting in circuit slowdown; EM can also cause permanent failure in circuits due to shorts or opens if the stress accumulates above a certain amount. As discussed in Chap. 1, conventionally EM is addressed usually by design rules (e.g., minimum metal width constraints) during the physical design phase and signoff [4], but conservative oversizing the metal wires can lead to significant area, power, and routing overheads. In addition, dynamic workloads and changing operating conditions can still lead to large variations in current densities, which can cause major EM threats at run time [5].

Similar to BTI, one important transient behavior of EM wearout is the recovery aspect, which refers to the stress relaxation in the metal line when the current is switched off as shown in Fig. 3.1. This can be referred to as a *passive recovery* condition not unlike in the BTI passive recovery case. With this EM passive recovery, the effect of the electron wind-induced stress can be relieved to only certain levels as demonstrated experimentally in [6, 7], but it cannot lead to full recovery due to the existence of a permanent unrecoverable component. In this chapter, we demonstrate

© Springer Nature Switzerland AG 2020
X. Guo, Mircea R. Stan, *Circadian Rhythms for Future Resilient Electronic Systems*, https://doi.org/10.1007/978-3-030-20051-0_3

Fig. 3.1 Illustration of EM
recovery: stress relaxation
occurs when the current is
switched off, which is similar
to the passive recovery
condition for BTI

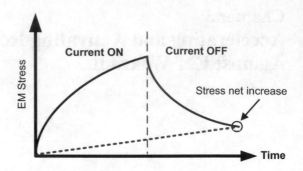

several solutions that can activate and accelerate the recovery of EM. We describe
experimental results on a set of on-chip metal lines on a test chip; the resulting
active accelerated EM recovery solutions are inspired by the notion of "reversing
the direction of wearout" as in the BTI case, except that the stress in the EM case is
current instead of voltage. We show that EM wearout also has a frequency dependent
behavior where the amount of permanent wearout depends on the periods of stress
and recovery (with the same duty cycle) again similar to what we found out for
BTI. By inserting the accelerated and active EM recovery periods periodically, the
overall mean time to failure (MTTF) of a metal line can be significantly extended.
Since EM active recovery happens when the current flows across the power grid,
the current direction can be switched periodically such that power is delivered to the
logic in a seamless fashion, and the circuit doesn't need to be switched off. This can
lead to minimal performance overhead if implemented with the necessary circuit
assist infrastructure. This chapter will present the experimental setup, theory, and
measurement results for the accelerated and active EM recovery techniques.

3.2 EM Wearout and Recovery Mechanisms

Compared to BTI wearout, the physics of EM has been less controversial. It is
widely accepted that EM is the result of momentum transfer from the electrons,
which move in the applied electric field, to the ions which make up the lattice of the
interconnect material. Figure 3.2 illustrates the process. Current flowing through a
metal line produces two forces, the first one is electrostatic force F_{field} caused by
the electric field strength in the metallic interconnect; this force is usually small and
can be ignored. A second force F_{wind} is caused by the momentum transfer between
electrons and metal ions in the crystal lattice; this force acts in the direction of
current flow and is the main EM source [4]. Over time, this can lead to resistance
increases and eventually to open or short circuits.

The EM recovery (healing effect) is caused by the atomic flow in the opposite
direction to the electron wind force F_{wind} during or after EM stress. This back-flow
of mass begins to take place once a redistribution of mass starts to form. The cause

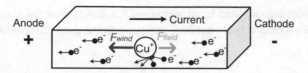

Fig. 3.2 Electromigration mechanism: EM is the result of the dominant force F_{wind}, that is, the momentum transfer from the electrons which move in the applied electric field

Fig. 3.3 Electromigration stress and passive recovery: EM mainly affects the power delivery network (PDN). EM healing occurs when current is removed, but the recovery is partial and slow

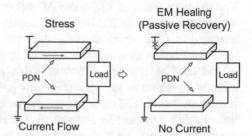

of this back-flow of mass is diffusion due to in-homogeneity, such as temperature and/or concentration gradients, resulting from the EM damage [8]. This passive recovery tends to reduce the failure rate during EM and partially heals the damage after current is removed. Due to this effect, the signal metal lines suffer fewer EM failures because most signal interconnects are stressed under bidirectional currents which correspond to the charging and discharging processes. Unfortunately for the power delivery network (especially the global PDN) shown in Fig. 3.3, EM doesn't have the same luxury because of the unidirectional current flow, and this can cause huge power net IR drop and permanent failures. Thus, in this work, we mainly focus on the EM issues that occur in the power delivery network.

The effect of EM is usually characterized by the mean time to failure (MTTF). The MTTF of a single metal interconnect caused by EM is given by the well-known Black's equation [9]:

$$\text{MTTF} = \frac{A}{J^n} exp\left(\frac{E_a}{kT}\right) \tag{3.1}$$

where A is a constant depending on the cross-sectional area of the interconnect, J is the current density, n is a scaling parameter (equals to 2 for void-nucleation-limited failure and 1 for void-growth-limited failure [10]), T is the temperature in $Kelvin$, k is the Boltzmann constant, and E_a is the activation energy. Equation (3.1) shows that the current density J and the temperature T are deciding factors that affect MTTF due to EM.

Recent EM modeling frameworks [11, 12] have shown that EM wearout experiences several phases. The first phase is called void nucleation during which the stress builds up until it reaches the critical value $\sigma = \sigma_{critical}$; the resistance during this phase is almost unchanged. Following the void nucleation phase, the generated voids start growing and lead to an increased resistance over time; as a result, the PDN becomes a time-varying network and the voltage drop changes over time.

3.3 Prior Work on EM Recovery

The recovery effect of EM under AC stress was studied in [13]; the experimental results showed that the EM lifetime increases with frequency. This effect was further analyzed in [14], which demonstrated that the healing can increase the lifetime by several orders of magnitude depending on the metal used. In [15] the authors looked into EM recovery in TSVs of 3DIC. While [6] suggested that EM is not fully recovered even under an opposite polarity pulse current with 50% of duty cycle, this just means that the EM stress and passive recovery are not symmetric, and there is an irreversible component for EM as well. The modeling work [12, 16] used theoretical physics to suggest that high temperatures can speed up atom diffusion towards the cathode and lead to higher rates and levels of recovery; but these works were solely based on simulations with no experimental validation. Ohfuji and Tsukada [17] were the first to conduct EM recovery experiments under different temperatures, but the goal of that work was to understand the transient resistance change during a temperature sweep up to 400 K, and a first-order model was proposed to capture this behavior. In [18], the authors demonstrated that at low frequency, when the current flow is interrupted, the stress gradient is sufficient to effectively counter the effect of EM and allow stress relaxation and consequently longer lifetimes. But this was still under passive recovery where the current is turned off in pulsed DC (PDC) operations. In our work, we explore the impact of both temperature and current direction on EM recovery. We also explore the irreversible component of EM wearout and its frequency dependence behavior and show that through several "deep healing" methods, EM-induced MTTF can be significantly extended.

3.4 "Reversing" the Direction of EM Wearout

In this section, we show how to achieve active accelerated EM recovery. During a recovery period, instead of relying solely on passive recovery, several recovery boosting techniques can be applied, and they are shown in Fig. 3.4; condition 2 shows the active recovery case where the direction of current is reversed to activate and assist the metal ion back-flow. In the previous chapter, we discussed that high temperatures are able to increase the kinetic energy for charge carriers. More importantly, increased temperatures can lead to recrystallization accompanying an increase in grain size and defect decay, such as the annihilation of vacancies at film surfaces [17]. Overall, the EM-induced damage and stress can be healed and relaxed using high temperatures, and this is shown as case 3. Case 4 illustrates the combined condition of 2 and 3. To validate these recovery conditions, we also

Fig. 3.4 Electromigration
"reversing" techniques: 0
shows the normal operation
condition, 1 refers to the
passive recovery, and 2, 3,
and 4 are proposed active and
accelerated recovery solutions

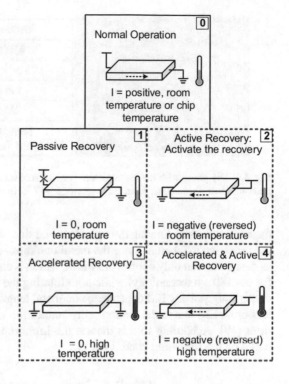

conduct actual hardware testing on a set of on-chip metal lines. Details of the setup
are presented in the next section. We also provide a comprehensive study on EM
frequency dependent behaviors with the same setup.

3.5 Test Setup

3.5.1 Test Structure

Since EM mainly happens to on-chip metal wires, and there are no commercial test
infrastructures which contain only on-chip metal lines, we fabricated a test chip in
180 nm bulk CMOS technology using dual-damascene Cu interconnect. The test
structures are a series of "long" and "narrow" on-chip metal lines. While this is a
mature technology we believe that most of the conclusions will also be qualitatively
valid for more recent advanced technology nodes since the fabrication methods and
materials are pretty much the same. Figure 3.5 shows the die photo along with
the dimensions of the metal wires. The metal wires are fabricated in the highest
metal layer (M6) of the technology since this metal layer is more likely being
used in the global power delivery network. The resistance change ΔR over time
is measured during stress and recovery phases. Based on Eq. (3.1), EM depends on

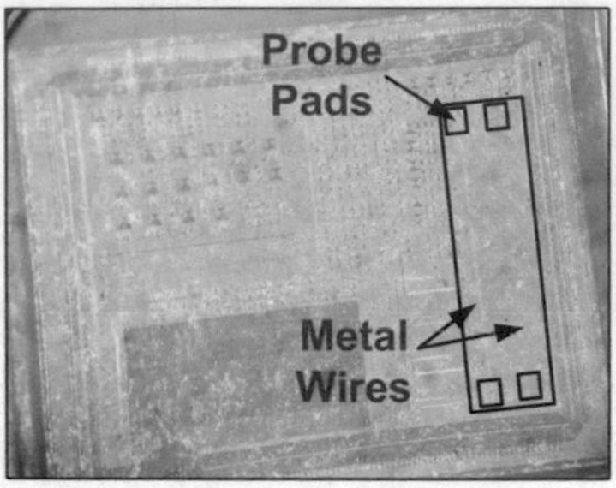

Technology	180nm
Material	Copper
Thickness	0.8um
Length	2.673mm
Width	1.57um
Resistance (@Room Temperature)	35.76Ω

Fig. 3.5 Die photo with the test structure for EM recovery: on-chip "long" and "narrow" metal lines and their dimensions; room temperature is ~27 °C in our test case

the current density flowing through the metal line, which is inversely proportional to the cross-sectional area. Since the metal thickness depends on the process itself we, as designers, can only control the width, which is chosen "narrow" enough (1.57 μm for the 180 nm technology) while not violating the design rules. Although Eq. (3.1) doesn't suggest any EM dependence on metal length, shorter metal wires experience less or no EM due to the "immortality" condition which is also known as the Blech limit [19]. According to this theory, the immortal metal segments can be filtered based on the following condition:

$$(j \times l) \leqslant (j \times l)_{critical} = \frac{\Omega \sigma_{critical}}{eZ\rho} \qquad (3.2)$$

where l is the metal length, Ω is the atomic volume, e is the electron charge, eZ is the effective charge of the migrating atoms, ρ is the wire electrical resistivity, and $\sigma_{critical}$ is the critical stress required for the failure precursor nucleation (void/hillock). Equation (3.2) means that the length of the metal line needs to be picked long enough to capture the EM dependence. As shown in Fig. 3.5, our metal line test structure is across the whole die area, and the length is 2.633 mm. Probe pads are placed on each end of the metal line, and they are used for connecting with bonding wire to communicate with the external measurement board; thus, they can be probed directly.

3.5.2 Measurement Setup

In our measurement setup, we first bond the wire-under-test to a regular dual in-line (DIP) package. Compared to the on-chip metal wires, the bonding wires employ a much larger cross-sectional area (>10×), and are believed to be much less impacted by EM; thus, the overall EM effects will be dominated by the on-chip metal wire as desired. Figure 3.6 shows the whole measurement setup, where the bonded chip is connected with a constant current supply using high temperature wires (red and

Fig. 3.6 Electromigration stress and recovery measurement setup

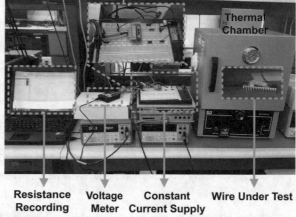

Resistance Recording Voltage Meter Constant Current Supply Wire Under Test

black wires shown in the figure) on a breadboard. In this way, only the wire-under-test is exposed to the high temperature environment. A voltmeter (Analog Discovery from Digilent [20]) is connected in parallel to the test structure for measuring the voltage drop ($\Delta V(t)$) on the wire-under-test. The voltmeter is USB-powered and supports visualization and data recording. We sample the voltage every minute and save it in a file for further processing. The resistance change over time due to EM can be calculated as $\Delta R(t) = \Delta V(t)/I$ based on the Ohm's law, where I is the constant current fed into the wire. The current value I and stress temperature T are decided based on both Eq. (3.1) and trial experiments so that MTTF is estimated to be in the range of few days. Before the stress phase, we first wait for a reasonable amount of time to ensure that a steady-state temperature was reached by the thermal chamber which allows a temperature fluctuation of $\pm 0.3\,°C$ only. The current supply is able to provide bidirectional DC current, and this allows a short switching time (less than 2 s) between the stress and recovery phases.

3.5.3 Test Cases

All accelerated stress tests are conducted on "fresh" test chips with the same test structures. Table 3.1 summarizes all test cases, where "EMST" stands for accelerated stress conditions, "EMPR" refers to passive recovery, and "EMASH" represents accelerated and active recovery conditions. The measurement halts when EM breakdown occurs, for example, in case EMST3, which provides a reference on how long it takes to reach the metal breakdown point (TTF). The case EMST1 is used as a baseline stress condition for comparisons. Similarly, the case EMPR1 is used as a baseline recovery condition for comparing against the accelerated and active recovery conditions (EMASH1 and EMASH2). EMST4 and EMST5 are test cases where metal lines are stressed under the same accelerated conditions

Table 3.1 Test cases for EM stress and recovery[a]

Condition	Case index	Wire no.	T (°C)	J (MA/cm^2)	Time (h)	Comments
Stress (active)	EMST1	1	27	7.96	12	Baseline for stress
	EMST2	2	230	7.96	12	–
	EMST3	3	230	7.96	20	Metal broke
	EMST4	4	230	7.96	12	–
	EMST5	5	230	7.96	6.7	–
Recovery (sleep)	EMPR1	2	27	0	20	Baseline for recovery
	EMASH1	4	230	−7.96	10	–
	EMASH2	5	230	−7.96	10	–

[a]*ST* accelerated stress test, *PR* passive recovery, *ASH* accelerated active recovery

except that EMST4 is stressed with a shorter period. These two stress periods are followed by two accelerated and active recovery periods (EMASH1 and EMASH2), respectively.

3.6 Experimental Results for EM Active and Accelerated Recovery

This section presents the experimental results from test cases summarized in Table 3.1. Figure 3.7 shows the measured EM-induced resistance change under accelerated stress and recovery conditions with very high constant current density ($\pm 7.96\,\mathrm{MA/cm^2}$) and elevated temperature (230 °C). During the accelerated stress phase, the results indicate that the EM evolution consists of two distinct phases as has been discussed in Sect. 3.2—the void nucleation phase and the void growth phase. During the void nucleation phase, the EM-induced stress increases until it hits a critical value, when voids are generated; before this point, the resistance stays almost unchanged. Following the void nucleation phase, these generated voids start growing and lead to an increased resistance over time. Our experimental results agree with measured data in [13, 21], and are also consistent with what has been predicted by recently proposed physics-based EM models [16, 22].

During the accelerated and active recovery phase, a reverse current (with the same absolute value as in the stress phase) and elevated temperature are applied; Fig. 3.7 demonstrates a case where the active recovery is much faster than passive recovery, and more than 75% of EM wearout can be recovered within only 1/5 of the stress time. Figure 3.7 also shows the results for the passive recovery case as a baseline for comparisons, where recovery saturates to a very high-resistance value after a short period of true recovery; this saturation continues even after extended recovery periods.

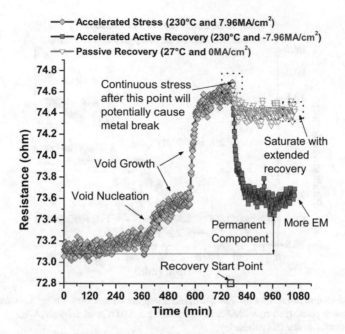

Fig. 3.7 Measurement results for EM degradation and recovery under passive recovery (Fig. 3.4 no. 1 or test case EMPR1 in Table 3.1) and proposed recovery conditions (Fig. 3.4 no. 4 or test case EMASH1 in Table 3.1, at 230 °C and ±7.96 MA/cm²) during the void growth phase: there is still a permanent component even under accelerated and active recovery

However, our results also suggest that there is still a lingering permanent component in accelerated and active recovery case, which is a similar behavior to what we found in BTI wearout measurements. This inspires us to explore the frequency dependence which is inspired by the method we used in the case of BTI recovery. This suggests that applying in-time recovery for EM should lead to a similar reducing, avoiding, or even eliminating the permanent component of EM; measurement results shown in Fig. 3.8 demonstrate exactly this. The results show that by scheduling the recovery intervals in the early phase of void growth, EM can be almost fully recovered. The potential issue of scheduling recovery during void growth is that during recovery, there is still (reverse) current flowing through the metal, and this could lead to potential EM in the opposite direction as shown in the figure; this can thus add uncertainties in terms of the ultimate effect of stress and recovery. A more "economic" way is to schedule the recovery following some patterns of "circadian rhythms" even before voids nucleation occurs; measurement results of this strategy are shown in Fig. 3.9, where multiple short accelerated and active recovery intervals (30 min) are inserted in the very early phase of EM stress evolution (first 600 min). This results in a delay of void nucleation for a significant amount of time (almost 3× slower compared to Fig. 3.7). By employing such circadian rhythm-like scheduled recovery, the overall time-to-failure (TTF) can be

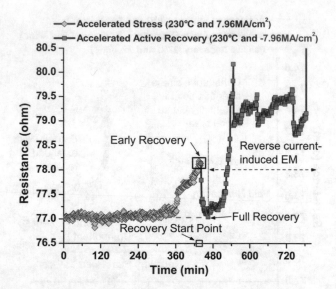

Fig. 3.8 Measurement results for EM accelerated and active recovery during the early period of the void growth phase (test case EMASH2 in Table 3.1 at 230 °C and ±7.96 MA/cm^2): almost EM is fully recovered under this schedule

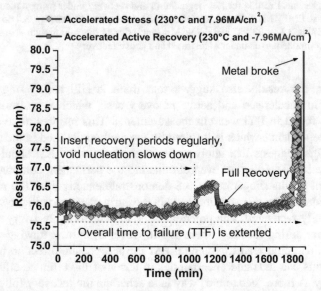

Fig. 3.9 Measurement results for proactively scheduled small periodic recovery intervals during void nucleation phase: it takes much longer for voids to nucleate, and the overall TTF is extended significantly

significantly extended. For example, Fig. 3.9 shows that TTF is increased by more than 2× with only five short recovery periods (2.5 h in total) scheduled in the early lifetime. If such a pattern is repeated further, we believe that TTF can be further extended.

Based on the above extensive accelerated stress and recovery tests, we conclude that EM recovery (back-flow) can be further activated and accelerated significantly. A stress/recovery pattern following an internal EM circadian rhythm has the potential to fully eliminate the permanent EM component. The ideal strategy is to insert short duration of EM accelerated and active recovery before any signs of void nucleation and repeat the pattern, the overall lifetime of the on-chip metal wire can be extended significantly. The resistance thus stays almost unchanged over the whole lifetime span. It is worth mentioning that BTI active recovery needs to happen when the transistors are switched OFF, while EM active recovery can actually happen during an ON period (with the right circuitry in place) with a reverse current flowing in the metal while the circuit is still powered on; this can lead to opportunities of scheduling BTI and EM recovery in an interchangeable way without introducing significant performance overheads. Details of such an implementation, including how to determine an optimal periodic rhythm, will be discussed in the following chapters.

3.7 EM Signoff Considering Accelerated and Active Recovery

EM issues are usually addressed by back-end-of-line (BEOL) design rules or/and by adding design margins. In the first method, current rule limits in metal wires are set by foundry to ensure reliable operations over a prescribed time period without significant EM damage. Designers need to follow these rules when doing floorplanning and physical design. For example, the number of power straps, number of VIAs, and PADs will be decided partially by these EM design rules. EM signoff tools (such as Cadence Voltus [23]) can also be employed to analyze and optimize the full-chip EM and IR drop. In the second method, designers need to assign larger margins to guardband against EM-induced delay degradation similar to what was done for BTI wearout—this results in a design and sign off with either a shorter lifetime or slower speed [24]. This margin could also be a voltage margin which is added when the resistance of the PDN mesh increases due to EM to keep the drive current the same so as to achieve the same performance. In summary, both methods require extensive design-time estimations and can lead to overhead for several metrics, especially performance.

With the accelerated and active recovery behaviors explored in this chapter, we show that the effect of EM stress can be fully recovered under certain conditions; thus, the EM signoff requirements can be relaxed significantly. This section discusses three different scenarios that can take advantage of the EM recovery properties.

3.7.1   Relaxing the EM Design Rules

Assuming that the EM recovery techniques have been implemented on-chip, then EM recovery can be accelerated and activated, and (almost) full recovery can be achieved.[1] For ease of comparison, we refer to the regular case without any proactive recovery as *Case EM_Reg*, and the case with recovery techniques (accelerated active self-healing) as *Case EM_ASH*. If the lifetime target remains the same and minimum metal width is being used, it means that the EM current design rule limits can be relaxed in the case EM_ASH since in-time proactive recovery can always bring the "aged" metal wire back to almost the original state; thus, this will result in fewer required power straps and less routing congestion as illustrated in Fig. 3.10. To quantify how much this current requirement can be relaxed (calculate the value of x), we use an industry-standard 28 nm FDSOI (fully depleted SOI) technology as an example. Since this is a relatively advanced node that has been deployed in many products, we believe that this analysis is representative enough to cover real design scenarios for most modern electronic chips. Listed in Table 3.2 is the maximum DC current allowed for this technology at a junction temperature of 125 °C for the top two metal layers that are usually used for power mesh routing. The EM current rules assure reliable operations of 10 years. Since the current limit is given at 125 °C, I_{dc} can be corrected by a temperature derating factor γ_{der} that is given in Table 3.3 for several temperatures. There is also a width derating factor that determines the maximum current, which can be relaxed by using a wider metal. Since in our case we assume the minimum width is used, this width derating factor is 1. Overall, the maximum DC current required by EM rules at temperature T is given by:

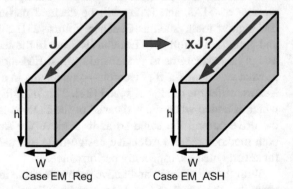

Fig. 3.10 Illustration of EM current rule relaxation due to recovery: current density requirement can be potentially relaxed, x is a number greater than 1

[1]Implementation details for EM on-chip recovery circuitry will be discussed in Chap. 4. The overhead and trade-offs when adding this type of circuitry will be discussed in Chap. 5. The main idea is that, by using switches, the direction of the current in the main PDN can be reversed as necessary, while the circuit itself remains powered on no matter the direction of the current in the PDN. For this section we assume that these have been implemented, and we only focus on studying the potential benefits enabled by EM accelerated self-healing techniques.

Table 3.2 EM line current limits @125 °C for 10 years of operations

Metal level	I_{dc} (mA)	Minimum width (μm)	Thickness (μm)
M10 (highest level)	0.408	0.4	0.88
M9	0.408	0.4	0.88

Table 3.3 Temperature derating factor $\gamma_{der}(T)$ and EM line current limit

T (°C)	$\gamma_{der}(T)$	$I_{dc}(T)|EM$ (mA)
125	1	0.408
110	2.792	1.139
50	40.62	16.573
27	68.312	27.871

Table 3.4 Summary of parameters for Black's equation

Parameter	Value	Unit
k	1.38×10^{-23}	J/K
E_a	0.9	eV
n	1	–
A	1.35×10^5	–

$$I_{dc}(T)|EM = I_{dc}(125\,°C) \times \gamma_{der}(T) \tag{3.3}$$

where $I_{dc}(125\,°C)$ is listed in Table 3.2. The current limit at various temperature can be calculated based on Eq. (3.3) and is listed in Table 3.3.

Based on Eq. (3.1) and our accelerated testing results, we can get a first-order estimation of the stress and recovery under normal operation conditions (normal current density and operating temperature). Assume that in the normal case, the junction temperature is 50 °C, which is close to a regular CPU die temperature with cooling. The current for this temperature is given in Table 3.3. The parameters used in Black's equation are given by the technology, and they are listed in Table 3.4. We also assume that the same temperature and current (absolute value) are applied during both stress and recovery periods. The equivalent stress and recovery conditions under normal operating condition to Fig. 3.8 lead to a result that all the voids can accumulate for at most 2.05 years (t_{nuc}) until nucleation; during this period the wire resistance remains almost unchanged. After that the nucleation and growth periods start and can continue for at most 168 days. If the direction of the current is reversed so that the active recovery starts, it will take at most 140 days to fully recover back to the "fresh" state. This periodic rhythm-like pattern can be applied repeatedly; thus, the resistance will never reach the rapid void growth phase which can potentially lead to rapid end of life for the wire. During all this time the circuit itself will be powered on through the PDN switches and will function unaffected no matter what the current direction in the PDN is (details in Chap. 4). Such a strategy opens opportunities for relaxing the EM design rules significantly. An example is illustrated in Fig. 3.11. It shows two cases where the regular case is without recovery, and it follows the EM current density limit which is J; this will guarantee a 10 years lifetime. If we double this EM current limit and keep the

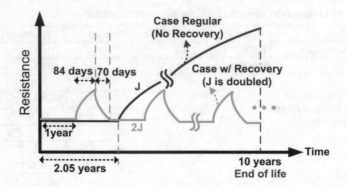

Fig. 3.11 Illustration of cases with and without recovery during normal operation: if the current limit is doubled, the metal wire ages faster (almost twice compared to before), but the in-time accelerated and active recovery can always bring it back to fresh state. Overall, the EM current limit can be relaxed while assuring the reliable operations during the lifetime span

Table 3.5 Estimated EM stress and recovery time under various operating conditions for 28 nm FDSOI technology

Description	Condition	Accumulation time	Nucleation time	Full recovery time	Comment
Accelerated condition[a]	$J = \pm\ 7.96\,\mathrm{MA/cm^2}$ $T = 230\,^\circ\mathrm{C}$	360 min	81 min	69 min	From measurement
Normal condition	$J = \pm\ 4.71\,\mathrm{MA/cm^2}$ $T = 50\,^\circ\mathrm{C}$	2.05 years	168 days	140 days	From calculation
Relaxed condition[b]	$J = \pm\ 9.42\,\mathrm{MA/cm^2}$ $T = 50\,^\circ\mathrm{C}$	1 year	84 days	70 days	From calculation

[a]This condition corresponds to the accelerated test condition in Fig. 3.8
[b]Under this condition, the EM current limit is doubled

temperature the same, the time to reach each EM stage will be about half according to the Black's equation; but increasing the absolute value of the current density also accelerates the recovery from EM when the current is reversed. This scenario can lead to potential "EM-free" operations over a long lifetime span. Details of the time length for different EM stress and recovery phases under various conditions are summarized in Table 3.5. Relaxed EM current rules will lead to fewer power straps during power network synthesis or reduce the need for metal layers which can potentially save fabrication costs with a smaller metal stack. It also offers the designer an additional flexible knob and can also bring a performance benefit and smaller margins; details will be discussed in the next sections.

3.7.2   Performance Improvement

Another potential benefit enabled by EM recovery can be the performance improvement due to less IR drop from power and ground mesh. As illustrated in Fig. 3.12,
when the resistance of the power/ground mesh (ΔR_{PG}) increases due to EM, the
performance of the circuit degrades. To achieve the same performance as before,
margins need to be added, but the power consumed by the mesh increases to match
the drive current. However, the introduction of EM recovery is able to fix the
effect of EM stress and mitigate the resistance increase as shown in Fig. 3.11 so
that the circuit always runs at a relatively high speed, the impact of EM-induced
IR drop becoming minimal. The potential margins and power overhead that are
wasted for matching the performance can be minimized as well. As the changing
load can lead to unpredictable dynamic stress conditions, and different metal layers
present different EM behaviors [2], proactively scheduling recovery periods can be
an effective candidate solution for improving the EM-induced performance loss.

3.7.3   Extending the Wire Lifetime

There are two ways of scheduling EM recovery proactively, the first one, inspired
by our experimental results shown in Fig. 3.9, is to insert a recovery period
even during the stress accumulation periods when resistance hasn't changed; the
second one, inspired by our measurement results shown in Fig. 3.8, is to recover
after a resistance increase has been detected and reached some predetermined
level (but still reversible). These two solutions are illustrated in Fig. 3.11. Both
solutions can lead to metal wires that get "refreshed" after each stress and recovery
cycle. Using these scenarios the lifetime of metal lines might no longer be a
bottleneck for systems that are expected to operate reliably for a long time. With
the recovery implemented on chip, the designers can still design and signoff the
chip in a traditional way, but this doesn't need to be driven by worst-case lifetime
considerations, which means the lifetime specifications for metal wire become a
reference instead of a constraint. Since recovery is not free as will be discussed
in the following chapters, the new trade-off becomes: how many recovery periods

Fig. 3.12 Illustration of
EM-induced IR drop on
power mesh

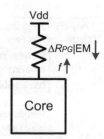

the system can afford vs. the overall lifetime/reliability budgets. The more and earlier the EM recovery duration is scheduled, the longer the metal line will last. This offers flexibility of "controlling" the lifetime at run time. For applications that are lifetime-bounded (e.g., automotive systems, medical implantable devices, etc.), recovery offers great opportunities of running for much longer with an extended lifetime for on-chip metal layers.

3.8 Summary: EM vs. BTI

Based on the experimental results and explored recovery behaviors presented in this chapter and Chap. 2, we summarize similarities and differences between EM and BTI behaviors in both stress and recovery in Table 3.6. During stress phases, both wearout mechanisms are accelerated by high temperatures and/or under high voltage/current stress. BTI is caused by voltage stress, and the device degradation is characterized by gradual threshold voltage (V_{th}) increase, while EM wearout happens due to current flow and it increases the metal resistance R, with the EM degradation behavior being very different from BTI. During the early phase of the EM stress period, the resistance stays almost unchanged, and this holds for a relatively long time until the void nucleation phase starts, the resistance presents a sudden increase and eventually reaches the failure state when the wire breaks and no current flows in the metal wire anymore.

In the recovery phase, we have demonstrated that high temperature and reverse stress (voltage for BTI, current for EM) can accelerate and activate the recovery for both phenomena. Even under these accelerated self-healing conditions, there can still be irreversible components which cannot be recovered within a reasonable period. We demonstrated that both BTI and EM present frequency dependence

Table 3.6 EM vs. BTI—similarities and differences during stress and recovery

Phase	Similarities	Differences
Stress (active)	High T accelerates both	BTI—voltage stress; EM—current stress
	High stress leads to more wearout	BTI—V_{th} increases gradually; EM—R doesn't increase until void growth and nucleation
		BTI—ΔV_{th} will saturate; EM—R increases (infinitely) until breakdown
Recovery (sleep)	High T accelerates both recovery	BTI active recovery—negative voltage; EM active recovery—reverse current
	Reverse stress activates both recovery	BTI recovery—device OFF; EM recovery—device ON (current flows)
	Irreversible part can be avoided	BTI recovery saturates; EM overrecovery can lead to more wearout in the reverse direction
	Frequency dependency behaviors	EM has a more complicated "circadian rhythm" due to complex stress behaviors

behaviors in which those irreversible parts can be eliminated by scheduling in-time circadian rhythm-like accelerated self-healing periods. Note that BTI recovery happens only when the circuit is in the OFF state, while EM recovery can be activated when circuits are ON with the help of switches that reverse the current flow through the wires while keeping the circuit powered on. Some circuit implementations that can enable this are presented in Chap. 4. A unique aspect of EM active recovery is that the reverse current can potentially lead to EM wearout in the reverse direction. This has been demonstrated by our experimental results shown in Fig. 3.8, where extended recovery period after full recovery causes more resistance increase. Thus this suggests that there is an upper limit for EM recovery time, and trying to overheal can result in undesired new EM issues. Because of this, EM needs to employ a more complex periodic rhythm compared to BTI. Careful considerations need to be taken when scheduling the accelerated and active recovery; more details are discussed in Chap. 5.

3.9 Conclusions

In this chapter, we mainly focused on recovery techniques for electromigration (EM) wearout that occurs to on-chip metal wires, especially to the power deliv-ery network. Figure 3.13 highlights the main contributions of this chapter. We demonstrated with extensive accelerated tests that EM recovery can be activated by reversing the current direction, and can be accelerated by raising the temperature. We showed several advantages that can be potentially offered by recovery with an example (in 28 nm FDSOI technology node). As EM effects can be reversed by proactive recovery periods, the EM design rules can be relaxed while still guaranteeing reliable operations and the reduced EM-induced IR drop can lead to less performance loss. For systems that require a long lifetime, the EM recovery techniques discussed in this chapter, if utilized properly, can offer opportunities to obtain long lifetimes in an economic way.

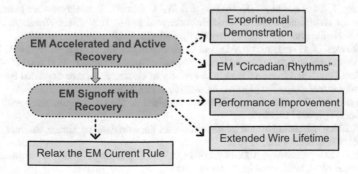

Fig. 3.13 Chapter 3 highlights

References

1. ITRS Report. http://www.itrs2.net/itrs-reports.html.
2. Gracieli Posser, Sachin S Sapatnekar, and Ricardo Reis. Analyzing the electromigration effects on different metal layers and different wire lengths. In *Electromigration Inside Logic Cells*, pages 93–98. Springer, 2017.
3. Bradley Geden. Understand and avoid electromigration (em) & ir-drop in custom ip blocks. *Synopsys White Paper*, pages 1–6, 2011.
4. Jens Lienig and Göran Jerke. Electromigration-aware physical design of integrated circuits. In *VLSI Design, 2005. 18th International Conference on*, pages 77–82. IEEE, 2005.
5. Taeyoung Kim, Xin Huang, Hai-Bao Chen, Valeriy Sukharev, and Sheldon X-D Tan. Learning-based dynamic reliability management for dark silicon processor considering em effects. In *Design, Automation & Test in Europe Conference & Exhibition (DATE), 2016*, pages 463–468. IEEE, 2016.
6. Ki-Don Lee. Electromigration recovery and short lead effect under bipolar-and unipolar-pulse current. In *Reliability Physics Symposium (IRPS), 2012 IEEE International*, pages 6B–3. IEEE, 2012.
7. MH Lin and AS Oates. Ac and pulsed-dc stress electromigration failure mechanisms in cu interconnects. In *Interconnect Technology Conference (IITC), 2013 IEEE International*, pages 1–3. IEEE, 2013.
8. Middle East Technical University Computer Simulation Laboratory (CSL):. http://www.csl.mete.metu.edu.tr/Electromigration/emig.htm.
9. James R Black. Electromigration—a brief survey and some recent results. *IEEE Transactions on Electron Devices*, 16(4):338–347, 1969.
10. Christine S Hau-Riege. An introduction to cu electromigration. *Microelectronics Reliability*, 44(2):195–205, 2004.
11. Xin Huang, Tan Yu, Valeriy Sukharev, and Sheldon X-D Tan. Physics-based electromigration assessment for power grid networks. In *Proceedings of the 51st Annual Design Automation Conference*, pages 1–6. ACM, 2014.
12. Xin Huang, Valeriy Sukharev, Taeyoung Kim, Haibao Chen, and Sheldon X-D Tan. Electromigration recovery modeling and analysis under time-dependent current and temperature stressing. In *2016 21st Asia and South Pacific Design Automation Conference (ASP-DAC)*, pages 244–249. IEEE, 2016.
13. Jiang Tao, Jone F Chen, Nathan W Cheung, and Chenming Hu. Modeling and characterization of electromigration failures under bidirectional current stress. *Electron Devices, IEEE Transactions on*, 43(5):800–808, 1996.
14. Jaume Abella and Xavier Vera. Electromigration for microarchitects. *ACM Computing Surveys (CSUR)*, 42(2):9, 2010.
15. S. Wang, T. Kim, Z. Sun, S. X. D. Tan, and M. B. Tahoori. Recovery-Aware Proactive TSV Repair for Electromigration Lifetime Enhancement in 3-D ICs. *IEEE Transactions on Very Large Scale Integration (VLSI) Systems*, PP(99):1–13, 2017.
16. V Sukharev, X Huang, and SX-D Tan. Electromigration induced stress evolution under alternate current and pulse current loads. *Journal of Applied Physics*, 118(3):034504, 2015.
17. Shin-ichi Ohfuji and Mitsuo Tsukada. Recovery of electric resistance degraded by electromigration. *Journal of applied physics*, 78(6):3769–3775, 1995.
18. IJ Ringler and JR Lloyd. Stress relaxation in pulsed dc electromigration measurements. *AIP Advances*, 6(9):095118, 2016.
19. Illan A Blech. Electromigration in thin aluminum films on titanium nitride. *Journal of Applied Physics*, 47(4):1203–1208, 1976.
20. Digilent Analog Discovery 100MS/s USB Oscilloscope & Logic Analyzer. https://reference.digilentinc.com/_media/analog_discovery:analog_discovery_rm.pdf.
21. Mircea R Stan and Paolo Re. Electromigration-aware design. In *Circuit Theory and Design, 2009. ECCTD 2009. European Conference on*, pages 786–789. IEEE, 2009.

22. Xin Huang, Valeriy Sukharev, Taeyoung Kim, and Sheldon X-D Tan. Dynamic electromigration modeling for transient stress evolution and recovery under time-dependent current and temperature stressing. *Integration, the VLSI Journal*, 2016.
23. Cadence Voltus IC Power Integrity Solution: Rapid power signoff and design closure. https://www.cadence.com/content/cadence-www/global/en_US/home/tools/digital-design-and-signoff/silicon-signoff/voltus-ic-power-integrity-solution.html.
24. Wei-Ting Chan, Andrew B Kahng, and Siddhartha Nath. Methodology for electromigration signoff in the presence of adaptive voltage scaling. In *System Level Interconnect Prediction (SLIP), 2014 ACM/IEEE International Workshop on*, pages 1–7. IEEE, 2014.

Part III
Implementing Self-healing on Chip

Chapter 4
Circuit Techniques for BTI and EM Accelerated and Active Recovery

4.1 Overview

In Chaps. 2 and 3 we mainly presented experimental demonstrations of accelerated and active self-healing techniques for BTI and EM, respectively. We also showed that these unique recovery behaviors can result in significant benefits for future resilient digital systems; but since design is always about trade-offs, these techniques do not come for free. Thus, it is necessary to explore how to utilize these recovery behaviors in an efficient and economical way. This chapter aims to answer the following research questions from a circuit design perspective:

- How to fully take advantage of the accelerated self-healing techniques on-chip for both BTI and EM wearout?
- What are the necessary circuit-level components (such as sensors) and assist circuitry for enabling a true accelerated self-healing system?
- What are the overheads introduced by the recovery circuits?

Figure 4.1 previews this chapter. As BTI and EM occur in different parts of the circuit, we introduce separate solutions for each of them. The chapter will begin by discussing the circuit implementation for accelerating and activating BTI recovery. Since BTI active recovery needs negative voltages, an on-chip, charge-pump based, negative voltage generator is first designed and simulated in a 28 nm FD-SOI technology. The area of the generator is estimated to be about only $300\,\mu m^2$, with a ripple of about 1.45%. Since BTI recovery occurs during "sleep," any leakage is going to represent a big concern. Power gating techniques [1] have been widely used to "shut off" the leakage paths from logic blocks by inserting power switches (also called sleep transistors) as headers, footers, or both. On top of the existing power gating infrastructure, BTI active recovery can be enabled by adding some extra logic so that the overall area overhead is minimal. Since high temperature accelerates recovery for both BTI and EM, a tiny on-chip heater design that is inspired by Weber et al. [2] will also be discussed. The heat generators should be enabled only

© Springer Nature Switzerland AG 2020
X. Guo, Mircea R. Stan, *Circadian Rhythms for Future Resilient Electronic Systems*, https://doi.org/10.1007/978-3-030-20051-0_4

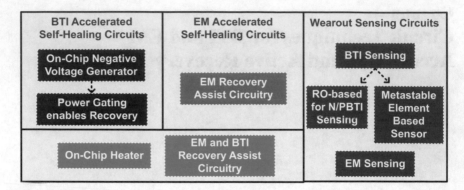

Fig. 4.1 An overview of the recovery circuit IPs in Chap. 4

during the recovery intervals, thus the overall power overhead can be compensated by reliability gains. For the EM accelerated and active recovery implementations, we introduce a multi-mode assist circuit scheme that is able to reverse the current direction without affecting the functionality of the circuit. The circuit scheme also supports BTI active recovery mode, during which transistors are reverse biased.

On-chip wearout sensors act as monitors for both the conventional adaptive solutions and the recovery solutions featured in this book, by translating device degradation into metrics that higher levels in the system stack can understand. The accuracy and reliability of these sensors is thus crucial for the system level management units [3] to use their information to monitor both wearout and recovery at the circuit level. This chapter presents three different designs of wearout sensors. Two of them are suitable for tracking BTI wearout and recovery, and one is for sensing EM wearout and recovery. The first BTI sensor is ring-oscillator based, and it can continuously track both NBTI and PBTI; the second one is a small metastable-element-based BTI sensor that serves as a trigger/alert for BTI stress and recovery. These sensors potentially can be deployed across the chip and be reused in various systems to guide the proactively scheduled recovery. Design details, simulation results, and physical implementations of all these circuits are detailed in this chapter.

4.2 Circuit Solutions for Activating and Accelerating BTI Recovery

4.2.1 On-Chip Negative Voltage Generation

In Sect. 2.7.1, we demonstrated that negative voltages are able to significantly boost the BTI recovery, but usually there are no negative voltages available on-chip; a negative voltage generator is thus needed. In this section, a modified switch-capacitor (SC) charge-pump negative voltage generator that is based on [4] is designed and simulated in a 28 nm industry-standard FD-SOI technology. Shown on the right half of Fig. 4.2 is the schematic of the generator. It works as follows: during the first half of the charge-pump cycle, $clk1$ is "0" and $clk2$ is "1," and the flying capacitor $C1$ is charged to V_{dd}. In the second half of the cycle, $clk1$ is "1 " and $clk2$ is "0," which means that the positive terminal of $C1$ is connected to ground and the negative terminal to $Vout$, and $C1$ is in parallel with $C2$. The charge will be redistributed until it reaches a new equilibrium point. The value of V_{out} is determined by the capacitance values of $C1$ and $C2$ ($C1/C2$) and the clock frequency f.

Figure 4.3 shows the simulation results for the negative voltage generator. It shows that it outputs a stable voltage of -300.6 mV after a start-up time of 638 ns under a clock frequency of $f = 66.7$ MHz. One of the most efficient ways of

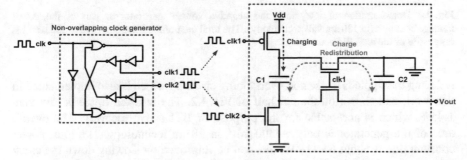

Fig. 4.2 A switch-cap-based negative voltage generator (designed in 28 nm FD-SOI technology) for delivering the negative voltage for BTI active recovery. $Vout$ is the negative voltage output

Fig. 4.3 A switch-cap-based negative voltage generator (designed in 28 nm FD-SOI technology) for delivering the negative voltage for BTI active recovery. $Vout$ is the negative voltage output

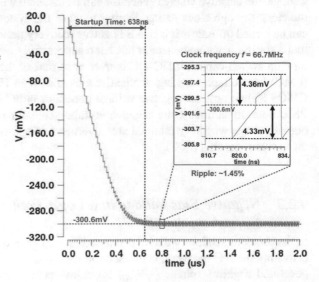

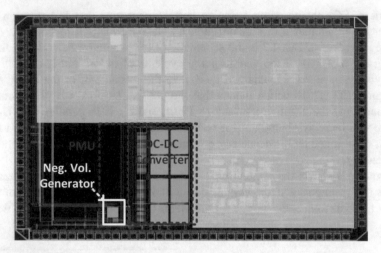

Fig. 4.4 Demonstration of integrating the negative voltage generator as part of the power management unit (in 130 nm bulk technology). The total area of the generator ($300 \times 315\,\mu\text{m}^2$) is only ~5% of the total PMU area

reducing the ripple is to use non-overlapping clocking which is also implemented in this work and shown on the left half of Fig. 4.2. The resulted ripple is less than 1.45%, which is acceptable for the purpose of BTI active recovery. The overall area of the generator is only ~4300 μm^2 in 28 nm technology. The total power consumption is about 64.47 μW; this can be improved by slowing down the clock frequency and sizing the switches. Since there are only four transistors in the generator, the introduced leakage power overhead is only 68.85 nW. For extra power savings, the negative voltage generator can be disabled when not used by turning off the clock through clock gating. During most of the system ON time, the generator can be turned off unless it is in a BTI active recovery period. It is worth mentioning that a similar negative generator has been successfully embedded in a multi-output on-chip switch-capacitor DC–DC converter as part of the power management unit (PMU) for voltage stacking applications as shown in Fig. 4.4. In a 130 nm bulk CMOS technology, the negative voltage generator only takes about 5% of the total PMU area. In summary, the negative voltage generator is able to provide a stable output voltage with very minimal area overhead compared to a regular PMU block on a system-on-chip (SoC).

4.2.2 Negative Bias Voltage in a Logic Path

In the previous sections we have discussed how to generate the negative voltage for activating the BTI recovery; in this section we discuss the feasibility of utilizing this generated negative voltage ($-V_{rec}$) for rejuvenating actual digital circuit blocks,

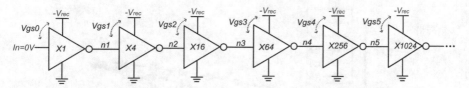

Fig. 4.5 A chain of inverter logic (FO4) with negative voltage supply $-V_{rec}$: when the supply voltage is negative, and the input of the first stage of the logic is "0," the gate and source voltage V_{gs} of PMOS transistors inside each logic stage is positive for all stages of logic so that BTI recovery can be activated. "nx" refers to the node voltage at node x, $n0$ is equal to 0 V

Fig. 4.6 An equivalent circuit for analyzing the behavior under active BTI recovery enabled by negative supply voltage $-V_{rec}$

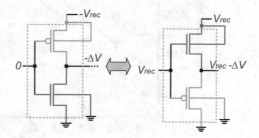

which usually include chains of logic gates. The questions to be answered in this section are if the negative voltage supply is able to rejuvenate all logic stages in the chain, and whether it introduces overheads such as leakage due to the added reverse bias across the transistors.

Figure 4.5 shows a chain of inverter logic with fanout of 4 (FO4); we use this as a proxy for the actual logic path in a functional block. In this example, the logic depth is only 6 to simplify the analysis, but in a real circuit IP block, it could be as high as hundreds or more [5]. During a BTI active recovery period, a negative voltage of $-V_{rec}$ is supplied to the logic path, the input V_{in} is set as "0," so PMOS transistors in each logic state experience a positive V_{gs} and BTI recovery can be activated, but not all transistors are biased at the same voltage $-V_{rec}$. In fact, the intermediate nodes (e.g., $n1$, $n2$, etc.) have different values due to the reasons explained next.

The left of Fig. 4.6 shows an inverter under negative voltage supply. To help the analysis, we add a voltage potential of V_{dd} to each node, thus an equivalent circuit is shown on the right half of figure; it is basically an inverter in which PMOS and NMOS transistors are flipped with an input equals to V_{dd}. The NMOS turns ON and the output node charges towards V_{rec}. But a voltage of ΔV is needed to keep the NMOS in the ON state—in super-threshold voltage region, this voltage is roughly V_{th}; in the sub-threshold voltage region, this delta voltage turns out to be lower than the threshold voltage. The functionality of this uncommon logic is that of a *weak non-inverting buffer* that passes weak "1" and "0." From the above analysis, the equivalent node voltage for the circuit on the left of Fig. 4.6 is equal to $-\Delta V$. As the signal propagates deeper in the chain, this delta voltage increases, thus the recovery voltage across the PMOS transistor could potentially reduce. To investigate what exactly this delta voltage is and how it affects each stage in a logic

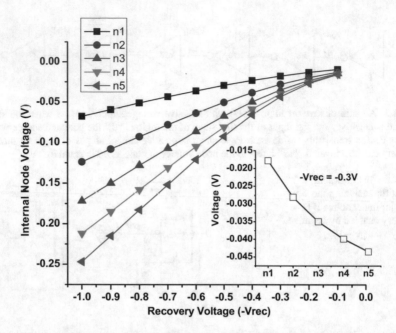

Fig. 4.7 Simulated internal node voltages of a chain of inverter logic during BTI active recovery with negative supply voltage. X-axis represents the applied voltage levels. As the logic depth increases, the node voltage level will increase slowly (an example of recovery voltage at -0.3 V shows this)

chain, we simulate the same circuit structure as shown in Fig. 4.2 in a 28 nm FD-SOI technology. Figure 4.7 presents the simulated results. The intermediate node voltages nx in each logic stage under different negative recovery voltage levels (in a range from $-V_{dd}$ to 0) are plotted in the figure. The result suggests that as the logic depth increases, the internal node voltage level increases under all recovery voltages. As an example, when $-V_{rec} = -0.3$ V, the node voltage drops from -0.0178 V (closer to "0") to -0.0435 V; this translates directly into a reduced bias voltage V_{gs} across PMOS transistors, and it is shown in Fig. 4.8. As we desire the same larger V_{gs} for each logic stage, the results show a reduction of this voltage, which seems thus to be suboptimal for the active recovery. But the good news is that this voltage doesn't decrease linearly, it reduces very slowly and actually saturates as shown in the figure. As a first-order estimation, it takes more than a hundred stages to reach "0" with a recovery voltage of -0.3 V. In summary, although applying a negative supply voltage to a whole function block doesn't mean that all PMOS transistors are under the exact same bias, they experience some level of positive V_{gs}, and BTI recovery can still be activated, as we have demonstrated that even a small positive V_{gs} can lead to a significant BTI recovery rate boosting. In addition, to compensate for this, the negative voltage can be delivered in a "fine-grained" way so that the PMOS transistors have a larger gate source bias voltage. A second solution is to

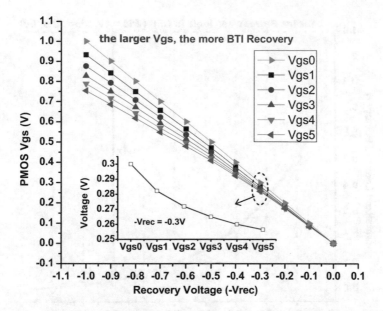

Fig. 4.8 PMOS transistor V_{gs} for each logic stage under different negative voltage supply levels. The larger V_{gs} is, the more negative bias PMOS transistors experience, the higher BTI recovery rate will be. As the logic goes deeper, V_{gs} decreases slowly. Depending on the logic depth, the negative voltage level needs to be carefully picked

adjust the negative voltage levels depending on the logic depth so that longer logic chains are able to be rejuvenated as fast as shorter chains.

Another important aspect that needs to be carefully evaluated when introducing a negative voltage supply in the logic is the power consumption. As the active recovery occurs when the transistors are OFF, the leakage power is the main metric of interest. As shown in Fig. 4.6, when under a negative supply voltage (active recovery), the PMOS transistor is in a forward bias (FBB) condition, under which V_{bs} is less than 0, and this can lead to a decreased V_{th} and potentially to an increased leakage current $I_{leakage}$. To explore this, simulations with the same setup as in Fig. 4.7 were conducted and the results are shown in Fig. 4.9. The baseline for comparison is the leakage power during passive recovery (V_{dd} is the nominal voltage, V_{in} is constant "1" or "0") and is used for normalization. All leakage power values ($I_{leakage} \times V_{rec}$) presented in the figure are normalized values. The results surprisingly show that the negative voltage doesn't introduce any leakage power overhead even when the negative voltage is biased at full range (-1 V in this case). Since we demonstrated with experiments that only a small negative voltage (such as -0.3 V) is good enough for boosting the recovery, the leakage power during sleep/recovery can actually be reduced significantly. Thus we conclude that it is feasible and safe to implement a negative voltage supply on-chip, and the recovery method is able to efficiently rejuvenate a circuit block from BTI without introducing additional leakage overhead.

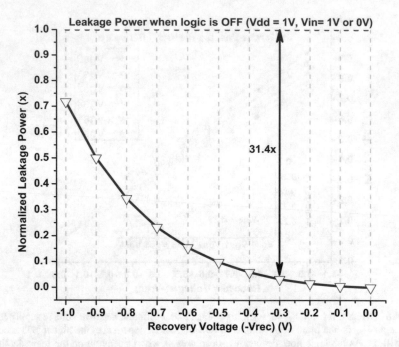

Fig. 4.9 Leakage power consumption during BTI active recovery. All values are normalized to the case when logic is OFF without any negative bias ($V_{dd} = nominal\ voltage$, $V_{in} = 0\ or\ V_{dd}$)

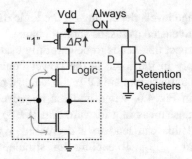

Fig. 4.10 Power gating structure: it can be used effectively to reduce leakage and help BTI recover passively, retention registers are used for saving the states. The voltage drop due to the resistance of header transistor ΔR can lead to performance loss; even worse, this transistor is ON most of the time and experiences BTI wearout as well

4.2.3   Wearout-Aware Power Gating

Power gating was first introduced as an effective method to reduce the standby leakage by inserting sleep transistors (ST) between the logic blocks and the actual power/ground rails [6]. Figure 4.10 shows a version of power gating with header sleep transistor inserted. Retention registers are used for storing the states temporarily when the logic is in sleep mode. Several works [6–9] have shown that the insertion of sleep transistors can also help with BTI-induced wearout since the structure enables a passive recovery mode while the circuit is in idle power-gated mode. However, power gating doesn't come for free. In addition to the area overhead, the sleep transistors also introduce an on-resistance ΔR, which further leads to a circuit delay penalty. The size of the sleep transistors together with the size of the power-gated blocks also determines the power-down or -up time [6]. When BTI wearout is considered, power gating design becomes more challenging. On one hand, power gating has a positive effect by protecting the logic against BTI wearout during idle, on the other hand, there is the negative effect of delay overhead. To address this trade-off, a clustered architecture was proposed in [6, 7] to leverage power, performance, and lifetime trade-offs by using two types of sleep transistors. In [10], a microarchitecture-level framework was proposed to mitigate wearout by utilizing power gating as a design knob for superscalar processors. A numerical model was proposed in [11] to analyze the potential benefits and limitations of power gating for reducing BTI. One aspect that is ignored in these previous works is that sleep transistors are ON most of the time, so they experience wearout the same way as the gated logic blocks. As a result, the circuit performance and lifetime can become even worse. This is shown in Fig. 4.11, where we simulate the threshold voltage increase (ΔV_{th}) of the sleep transistor vs. circuit performance loss in a 28 nm FD-SOI technology. The circuit consists of eight 49-stage ring oscillators running in parallel. This performance degradation can be as large as more than 10% under a small threshold voltage shift. To tackle this, Calimera et al. [7] proposed to realize NBTI-aware power gating by oversizing the sleep transistors, using

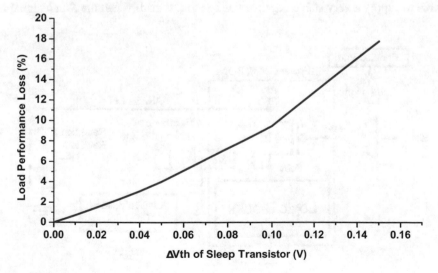

Fig. 4.11 Sleep transistor threshold voltage increase ΔV_{th} vs. circuit performance loss (8 49-ring oscillators running in parallel, the header transistor is sized as 1 μm wide)

forward bias to compensate and by reducing the stress time. Lee et al. [12] explored introducing redundancy in the sleep transistors (STs) to reduce the overall turned-on times of these transistors. Wu et al. [13] further analyzed the joint interdependent degradation effects on logic networks and sleep transistors by using redundant STs. In [14], two BTI-aware sleep transistor sizing algorithms were proposed to reduce the total width of STs based on the distributed ST network structure. Rossi et al. [15] performed an analysis and design flow to pick threshold voltage values for STs to optimize both leakage and lifetime.

In addition to these previous solutions, this section presents an orthogonal power gating solution that enables full BTI recovery by delivering the negative voltage supply to logic blocks. The structure also incorporates the active BTI recovery for the sleep transistor by using a higher-than-V_{dd} voltage; the proposed circuit topology is shown in Fig. 4.12. An NMOS switch is added to the virtual power supply node for delivering the negative voltage when the active recovery mode is enabled (Signal $Sleep = 1$). The capacitor $C2$ in the negative voltage generator (shown in Fig. 4.2) acts as a decoupling capacitor (Decap), so no external Decap for the negative voltage is needed. Figure 4.13 shows the functional simulation of the proposed power gating structure in an industry-standard 14 nm FinFET technology node. The load circuit consists of four 9-stage ring oscillators oscillating in parallel. The header transistor is sized so that the voltage drop is within 5% of nominal. V_{dd}_high is selected so that the negative bias across the PMOS header is ~ -0.3 V. The NMOS transistor is added to deliver the negative voltage of -0.3 V from the voltage generator. It is sized so that the switching times between modes are similar to the regular power gating case where there is no active recovery implementation. As shown in the figure, when the BTI active recovery mode starts, the structure is able to supply a very stable negative voltage to the load. When the $Sleep$ signal is

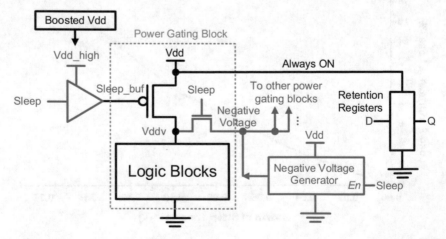

Fig. 4.12 An active recovery-enabled power gating structure: blue outline parts refer to the additional logic on top of the existing infrastructure. The Sleep signal is the trigger for starting the active recovery

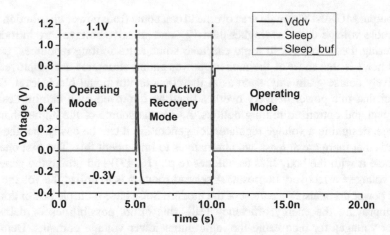

Fig. 4.13 Functional simulation of the proposed wearout-aware power gating structure in 14 nm bulk FinFET technology, signal names correspond to the ones in Fig. 4.12: when the *Sleep* signal is high, the BTI active recovery mode starts, the negative voltage is delivered as the virtual supply for the logic, the header sleep transistor is also in reverse bias mode because a higher-than-V_{dd} voltage is applied to the gate. The switching times between modes are similar to the ones in regular power gating structures

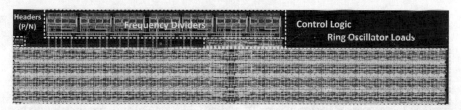

Fig. 4.14 The physical layout of the power gating that implements the BTI active recovery logic in 28 nm FD-SOI technology. The added overhead is only an NMOS header and some control logics. The load is 8 ring oscillators in parallel, and the frequency divider is used for off-chip frequency readout

set as "low," the circuit can quickly switch back to the normal operating mode from the active recovery mode. This switching time can be further reduced by optimizing the load size and the NMOS header transistor size.

The proposed wearout-aware power gating structure introduces some extra logic such as a buffer and an NMOS transistor, but the overhead of these components is minimal. For example, in the case shown in Fig. 4.13, the added area overhead is only 2.71 μm^2, which is close to area of an X5 flip-flop in the same technology. This can be seen in Fig. 4.14, which gives a physical implementation of the wearout-aware design. We integrate the BTI active recovery with the existing power gating structures. The load consists of 8 ring oscillators in parallel, and the frequency divider is used for off-chip frequency readout. The added logic includes an NMOS header (shown on the top left part) and control logic, which includes an input buffer and output MUXes. The main area overhead can come from power distribution since two more voltage domains (higher-than-V_{dd} and negative voltage) are introduced; delivering these voltages to logic can take some extra routing resources, but the good news is that most of the modern power gating structures are employed in a relatively coarse-grain way, such as at the core- or functional-block level. So we expect that this power delivery overhead can be improved by a careful designed floorplan and optimized routing options. As for the source of the higher-than-V_{dd} voltage, designing a voltage regulator for generating it can be costly, but the good news is that there are at least two alternatives to implement this. The first one is to combine it with the body bias techniques (e.g., [16, 17]) and utilize the generated high voltage for recovery purposes; A second solution is to utilize the voltage from other power domains. Since most of the modern SoCs have multiple power domains for improving the energy efficiency [18], this offers possibilities of delivering higher voltages for recovering the logic under lower voltage domains. Details of the implementations depend on the specific applications and design specifications. In summary, the proposed "recoverable" power gating design is orthogonal to the existing power gating structures, and it can be integrated with those existing structures and utilizing the same optimization framework for achieving ultra-robust digital systems.

4.2.4 On-Chip Heat Generation

As demonstrated in Chaps. 2 and 3, higher temperatures accelerate both BTI and EM recovery significantly; at the same time it is also well known that heat can degrade the processor performance. To address this conundrum, we propose to use small reconfigurable localized on-chip heaters for accelerating local recovery. These heater designs are inspired by previous work [2, 19, 20], where they were used as self-heating elements for thermal-aware testing on FPGAs. Thus, similar designs can be adapted to our high temperature-enabled accelerated recovery purposes. A proof-of-concept design is presented in Fig. 4.15. The theory behind it is to force high toggling rates of the reconfigurable ring oscillator to generate heat, and the output frequency can be selected based on the temperature requirement. The design is straightforward and easy to implement and control. *Accelerated recovery* signal serves as an enable signal and can be triggered when needed. Note that the length of the oscillator in terms of number of stages doesn't need to be always half of the total length—it depends on the desired recovery temperature.

Since each frequency corresponds to different temperatures, it is important to understand how these two parameters are related. Figure 4.16 shows a case where similar heating circuits are implemented on FPGAs, and the measured maximum temperatures during steady state are plotted [2]. Frequency and temperature have an almost linear relationship, where temperature increases by 10 °C with an increase in frequency by 100 MHz; heaters take some time to reach steady state due to relatively large thermal time constants. These times can be in the order of seconds or even

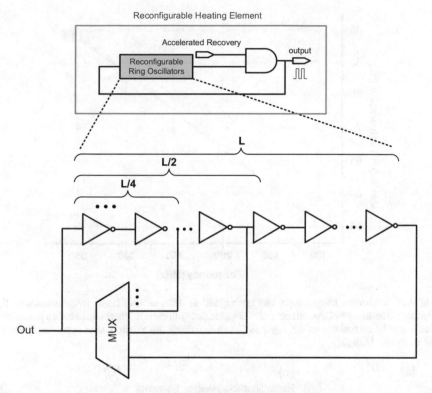

Fig. 4.15 Reconfigurable on-chip heating elements schematic: the output frequency of the ring oscillators can be reconfigured by selecting the length of the inverter chain (L, $L/2$, etc.). Different frequencies of the heating elements correspond to a wide range of temperatures

minutes, which we believe is still acceptable for recovery purposes. Authors in that work also observed that temperature swing with respect to idle temperature increases with higher frequency of the generator, and this swing can be as large as ±5 °C. On the other hand, our active recovery experiments results show that such a temperature swing has a relatively minor impact on recovery rate at high temperature values (>80 °C). Although the relationship in Fig. 4.16 might not be always the same for other designs or chips due to differences in thermal time constants, it still indicates that on-chip heaters are able to achieve a wide range of relatively steady high temperatures. A precalibration process will likely be necessary to determine the exact relationship for the target devices [20].

The distribution of the generated heat from the above discussed heaters across the desired region is determined by the placement of these heating elements. Ideally, a homogeneous temperature distribution across the block is desired for recovering all the logic circuits in that block. Here we discuss three different strategies of placing heaters across the chip. These are shown in Fig. 4.17. The first one is the most ideal case where the heating elements are spread evenly within a logic blocks so that

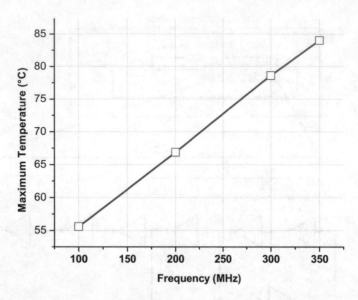

Fig. 4.16 Maximum temperatures that correspond to different oscillation frequencies with the heating elements on FPGAs, temperatures are sampled with external thermal sensors, a precalibration has to be carried out on the target FPGA to determine this relation, data is plotted based on measurement from [2]

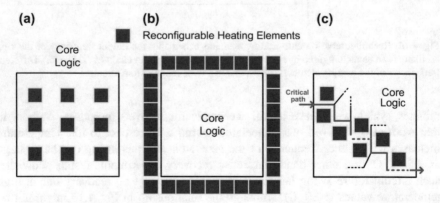

Fig. 4.17 Potential strategies of on-chip reconfigurable heaters: (**a**) evenly distribution; (**b**) ring placement; (**c**) critical path placement. Red squares refer to the heating element. An external controller is needed to configure the heater and select the blocks that need to be recovered

a homogeneous temperature distribution can be achieved across the whole area. In fact, such uniform placement can lead to a temperature profile where the edges of the block are at a slightly lower temperature than the center. To handle this, the heater in the center can be reconfigured to oscillate at a slower frequency and thus make the temperature across the whole area more uniform. The difficulty of implementing such a strategy is that it requires design efforts for place and route, especially for logic blocks that are not rectangular or square. A relatively easier way to place the heating elements is shown in Fig. 4.17b, where a "heater ring" is formed and placed surrounding the logic block. The advantage of this method is that the heating elements are separated from the logic, so they have less effect on performance during the run time. But the disadvantage is that the center of the logic block is not able to reach the desired recovery temperature due to the physical distances to the heaters. Thus this solution is applicable for those smaller blocks or those logic that are less wearout critical, and such heater rings can potentially be shared by logic blocks that are next to each other. As the required number of heating elements is proportional to the logic area, another, more economical way of placing heaters, is shown in Fig. 4.17c. The heaters are placed only close to critical paths which determine the overall performance of a logic block. With this setting, the number of heaters can be significantly reduced, and thus the area overhead can also be further reduced. The challenge is to have the right methodology to correctly identify critical paths and place heaters during the design phase. One possible solution could be to embed heating elements into scan chain cells and place them selectively along with the scan chain insertion during design for test (DFT). As many heating elements are expected on a single chip, a heating controller is required to configure the heaters individually; it also needs to activate/deactivate a subset of heating elements according to the desired maximum temperature. This controller can be a look-up-table (LUT), which takes the desired recovery temperature from the system as an input, and compares it to the actual chip temperature, which can be either read from built-in thermal sensors available on chip or predicted by use of thermal analysis tools, and then output the corresponding select signals for the right group of heaters. In summary, each of the above three placement strategies has some advantages and disadvantages. Careful design decisions that balance these trade-offs will be able to take full advantages of high temperature for effective self-healing.

As shown in Fig. 4.15, each reconfigurable heat element consists of a chain of inverters and a MUX. The length of the chain is decided by the desired frequency, which further depends on the desired recovery temperature. As an example, for an oscillating frequency in the GHz range (corresponding to $>80\,°C$), about ~41 stages of $X4$ inverters are required with a 28 nm FD-SOI technology; the total area of one such heater is around $16\,\mu m^2$, and the leakage power is about 16.8 nW. This is still much less than a regular on-chip temperature sensor (such as sensors proposed in [21, 22]). Thus it is feasible to put on-chip heating elements as their area and leakage overhead are acceptable. However, it should be noted that the on-chip heating elements are based on the assumptions that extra power consumption is within an acceptable range during recovery. For example, this power can be as large as mW depending on the target temperatures, which might not be suitable for all applications, especially for those with very tight energy efficiency requirements. But the trade-offs are between the energy and power consumption during OFF periods and the improved metrics during active time. Careful designs can potentially leverage these two so that the overall overhead is minimal and acceptable. An alternative and potentially more economical way of recovering with high temperature is to utilize the existing on-chip heat, which can be realized through core level redundancy or through active cores/elements that serve as "natural on-chip heaters" for the neighboring asleep cores. This will be further addressed in detail in Chap. 5.

4.3 Circuit Solutions for Activating and Accelerating EM Recovery

Our experimental results in Chap. 4 demonstrated that high temperatures accelerate recovery for on-chip metal lines from EM damage, while reversing the current direction is a way to further activate EM recovery. Since power rails suffer from single-direction DC current mostly [23, 24], we focus mainly on EM-induced effects in the power delivery network in this book. To implement EM active recovery with reversed current on chip, an assist circuit scheme that is able to support this without affecting the functionality of the load is presented in this section. The idea is inspired by several conceptual designs that were proposed in [25–27]; the key difference between our scheme and previous designs is the added recovery modes— our scheme is able to support both EM and BTI active recovery modes that were guided with the notion of "reversing" the directions of wearout. Besides, we also address the physical implementation and potential system-level integration solutions for the novel circuit scheme. Shown in Fig. 4.18a is the schematic of the proposed multi-mode EM/BTI recovery assist circuitry, which includes two layers of power gating, with VDD/VSS power grids in between. Functionally, it can be configured in three modes—*normal, EM active recovery*, and *BTI active recovery*. Figure 4.18b is the corresponding functional truth table—these three modes are illustrated in Fig. 4.19. Under the *normal* operating mode, the functional load operates similarly to a regular single-layer power-gated system; Under the *EM active recovery* mode, the direction of the current flowing through the VDD and VSS grid is reversed, but the current value still remains unchanged because of the symmetry of the scheme, thus the load (target circuit that is under the recovering power delivery network) still functions as under the *normal* mode. *BTI active recovery* is triggered when the load is in idle state, during which VDD and VSS nodes of the load are switched. This creates a BTI active recovery condition not unlike what has been described in Sect. 4.2.2. Depending on the input values, NBTI or PBTI active recovery can both be activated; this is shown in Fig. 4.18c.

To validate the functionality of the design, we implement and simulate the above recovery assist circuitry in a 28 nm FD-SOI technology. A set of ring oscillators running in parallel is used to mimic the load condition, the VDD/VSS grid is treated as a resistor whose values are picked based on reasonable PDN assumptions from the published literature. Figure 4.20 illustrates the functionality with simulation results under three different modes. Figure 4.20a plots the current flowing across the VDD grid during under *normal* mode and *EM active recovery mode*, respectively. As expected, the direction of the current is reversed, but the amplitude still remains

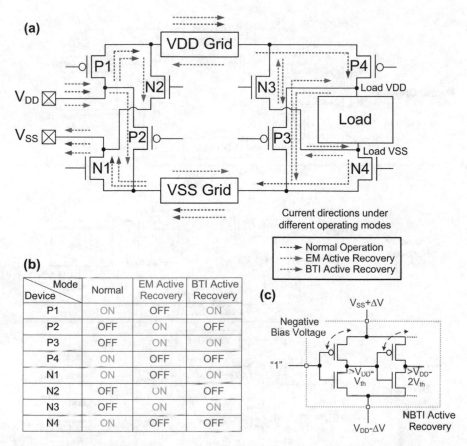

Fig. 4.18 Assist circuitry for activating BTI and EM recovery: (**a**) The main circuitry, arrows represent the current direction under different modes, V_{DD} and V_{SS} pins can be connected to the on-chip voltage regulator directly, or to the global power delivery network; (**b**) Truth table for three operating modes; (**c**) An example of activating NBTI recovery under *BTI active recovery* mode, for PBTI recovery, the input needs to be "0," ΔV represents voltage droop/increase or noise

the same; this ensures that the load operates in both modes without losing any performance. Under the *BTI active recovery* mode, the load is in sleep/idle mode, its VDD and VSS nodes are switched, and there is about a 0.2 V delta voltage droop/increase induced by the pass transistors; but even with this, the voltage is still large enough for activating BTI recovery (-0.816 V is much higher than -0.3 V demonstrated by our experiments in Chap. 2). Three modes are configured by a 3-bit controller, which is a decoder design based on the truth table in Fig. 4.18b.

One of the biggest design challenges for the active recovery assist circuitry is the voltage droop/increase at the load VDD/VSS nodes introduced by dual layers of header or footer transistors during *normal* operation and *EM active recovery* mode when load performance is critical. Another potential concern is the switching

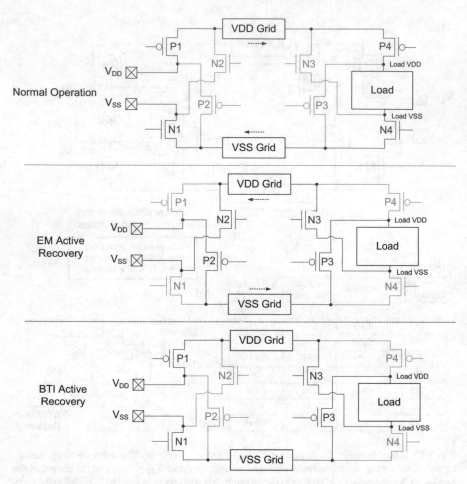

Fig. 4.19 Three modes of the assist circuitry for activating BTI and EM recovery: *normal operation*, *EM active recovery*, and *BTI active recovery*. VDD and VSS grid in the figure refer to power delivery network (PDN)

time (retention time) between modes. Since both metrics highly depend on the load capacitance, we now study how the size of the load affects the performance. Figure 4.21 demonstrates that by increasing load size, the oscillation frequency of the load ring oscillator degrades linearly because of the voltage drop/increase across the footer/header transistors. Switching time also increases with the increased load, but at a slower rate. To compensate this performance degradation, the header/footer transistors need to be upsized accordingly, and this is also true when only one layer of the power gating is used. This result also indicates that each load will have its own optimal design point which gives the optimal values in terms of area, performance, and other metrics. Thus, another careful design choice needs to be made to balance the trade-offs of performance vs. area vs. reliability.

Fig. 4.20 Functionality simulation for EM/BTI active recovery assist circuitry in 28 nm FD-SOI technology. (**a**) The current direction is reversed under *EM active recovery* mode, but the current value is still the same so that the load still runs at the same frequency. (**b**) Under *BTI active recovery* mode, load VDD and VSS values are switched so that the voltage across the transistors is reverse biased

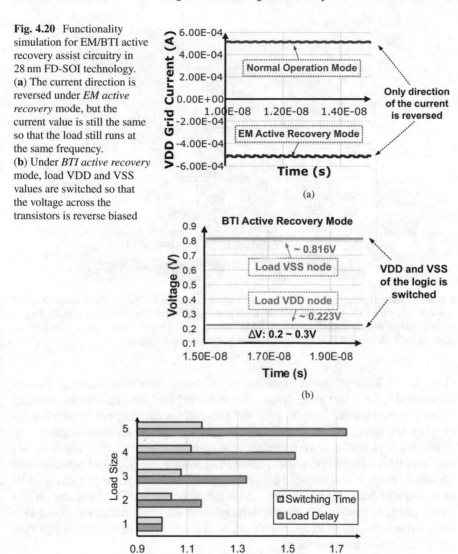

Fig. 4.21 Load size vs. performance and switching time: increasing the number of loads will reduce the performance and increase the switching time between modes, to compensate the degradation, header/footer transistors need to be upsized, which will further increase the area. This trade-off needs to be carefully considered during the design process

Figure 4.22 previews the physical implementation of the VDD grid with the integrated assist circuitry (10-layer metal stack is assumed similar to many advanced technology nodes such as 28 nm FD-SOI and 14 nm FinFET). The design has a global PDN grid which usually uses the top one or two metal layers that are wide and thick, thus being relatively robust against EM. Local VDD/GND grids that are

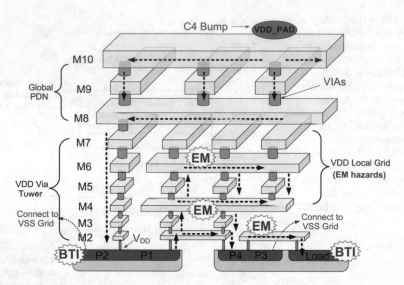

Fig. 4.22 Vertical cross section of the physical implementation for the EM/BTI active recovery assist circuitry for VDD grid (VSS grid can be implemented in a similar way): EM hazards happen at high current density regions, which could be caused by faster switching activities on the load logic; at the logic level, BTI hazards happen due to the continuous stress

closer to the logic use lower metal layers which are more EM sensitive; $P1$–$P4$ correspond to the power gating header transistors in Fig. 4.18a. This implementation which is able to protect the local grids and also enables the design of localized assist circuitry for individual loads suffers from EM wearout of different degrees. The structure is very similar to a conventional power-gated PDN, on top of which we add one more layer of header/footer transistors. The overall area overhead introduced by the assist circuitry is determined by design specifications in terms of maximum IR drop, and also depends on load size and switching behaviors. The good news is that power gating techniques are already widely used; the implementation shown here demonstrates the ease of integrating the assist circuitry with the existing design flow and power gating infrastructures.

4.4   BTI Sensing

As BTI wearout happens at the transistor level, and some of the recovery decisions or adaptive solutions occur at the circuit or even higher levels of a system hierarchy, on-chip wearout sensors have been an important component that bridges these various levels of abstraction in a cross-layer fashion. The sensors act as monitors for both the conventional adaptive solutions [28, 29] and the recovery solutions discussed in this book. The accuracy and reliability of these sensors will be crucial for the system level management units [3] to use their information to monitor both wearout and *recovery* at the circuit level. We can expect the total number of on-chip wearout sensors in many-core systems to reach thousands [3, 30], with hundreds of sensors per core. There have been various sensor designs recently that use advanced circuit techniques to achieve good sensitivity in the presence of variations. However, many of these are analog in nature, and are also not able to track *recovery*, thus this limits their wide deployment and reuse. In this section, we present two types of BTI sensing techniques: the first one is ring oscillator based and can differentiate NBTI and PBTI, thus the structure can be used for BTI testing and characterizing, the second one is a novel type of small and embeddable digital BTI wearout sensors that are based on metastable elements. These sensors are able to track both wearout and *recovery*, and are also able to detect BTI-induced critical path reranking [31, 32] by tracking multiple paths simultaneously. This section details the design, implementation, and placement methodology of the novel BTI sensors.

4.4.1 Previous BTI Sensing Techniques

The effect of BTI can be mapped to several metrics, the most straightforward one being frequency degradation—this is also the most common way of detecting BTI [33]. In [34–36], ring oscillator (RO) based sensors were proposed using tracking and reference ROs to detect the frequency difference. These sensors can achieve very fast tracking, but area and power overhead are significant. Additionally, intra-die variations of the ring oscillators will make the measurements inaccurate. Some work also proposed to use on-chip circuit such as SRAM to predict wearout [37]. Another metric that can be used for BTI sensing is leakage current or drain current—there is some work that uses highly accurate current testing techniques to track BTI [38, 39], but it is difficult to use these techniques for real-time on-chip monitoring; these circuits are usually complex and need off-chip measurements. In [40], a self-testing technique for tracking NBTI was proposed—this sensing technique is robust across PVT variations, but it is still ring-oscillator based and has large area overhead. Wooters et al. [41] proposed a small embedded NBTI sensor based on a metastable circuit. The circuit consists of a pair of cross-coupled inverters. Instead of providing data at every time point, these sensors can provide feedback for the system when the circuits have degraded by a certain critical percentage ΔI. This scheme allows smaller sensor sizes while still achieving good accuracy. Suresh and Burleson [42] extended the work by designing a fine-grained sensor that can track both N/PBTI; the resolution time measurement is done by using a ring oscillator-based time-to-digital converter. Alidash et al. [43] proposed an on-chip NBTI sensor which is able to isolate the NBTI and PBTI aging effects as well, but it requires multiple copies of the same sensor, and this could lead to large area overhead. Machine learning-based BTI sensing techniques appeared recently, the idea being to identify the delay degradation for a large pool of paths through monitoring the delay of a selected small subset of paths [44, 45]—as there is no requirement to equip every critical path with a BTI sensor, the area overhead can be reduced. In summary, all these previous works focused on tracking the degradation of BTI of critical path/block and did not really consider how to embed these sensors into a system and/or a design flow. Also to the best of our knowledge, there is no BTI sensor that has been designed for tracking proactive recovery, and tracking multiple (critical) paths simultaneously with one embedded sensing structure, thus if critical path reranking [31, 32] happens, the sensor information will not be accurate any more. In this section, we present two novel types of BTI sensors, in which the first type is able to separate NBTI and PBTI with only one test structure. The second type is able to support both stress and recovery sensing with an ability of tracking multiple (e.g., 4 or more) critical or potential critical paths simultaneously. The compatibility of the sensors makes them easy to embed as a single IP block with a conventional top-down ASIC design flow.

4.4.2   Ring Oscillator-Based Test Structures for Separating NBTI and PBTI

For older technology nodes PBTI was considered as a secondary effect compared to NBTI, but it has become a considerable reliability concern as high-k dielectric material and metal gate were adopted in modern process nodes for gate leakage reduction [46], thus sensing and characterizing PBTI and NBTI are both important. Conventional sensing techniques or test structures such as ring oscillators don't allow individual and independent measurement of PBTI and NBTI. Prior work [43, 47] proposed solutions of isolating NBTI and PBTI by including multiple copies of the sensing structure and configuring them for different modes. These solutions work effectively, but large area overheads are introduced due to the fact that multiple sensors might be required for each BTI mechanism in a large SoC. Here we present a simple and integrated ring oscillator-based sensing stage that can monitor NBTI and PBTI separately while not introducing significant area overhead. Figure 4.23a shows the schematic of one such stage, where the red inverter is the circuit of interest which experiences NBTI or PBTI. If the ring oscillator cells are replaced with this structure, the oscillation frequency degradation will be able to reflect either PBTI or NBTI depending on the stress mode. The transmission gate serves as a switch for stress mode (NBTI or PBTI) and test mode, where the oscillation frequency of the ring oscillator is sampled. Figure 4.23b is the functional truth table for the sensing stage. Under NBTI mode, the circuit is configured such that only the PMOS is under constant stress (full V_{dd} swing) in each stage; the PBTI mode works similarly. When under test mode, the sensing stage is configured as a regular single-stage inverter, and is sampled as a normal ring oscillator stage. The output frequency of the ring oscillator is sampled under this mode.

The design has been implemented in a 28 nm FD-SOI technology. Figure 4.24 is the layout for a 37-stage ring oscillator that is implemented with the sensing element discussed above. The overall area is about 92 μm^2, and it is only ~1.6× larger

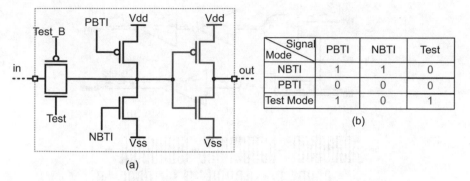

(a)

Signal Mode	PBTI	NBTI	Test
NBTI	1	1	0
PBTI	0	0	0
Test Mode	1	0	1

(b)

Fig. 4.23 Sensing element for separating NBTI and PBTI. (**a**) Functional schematic of 1 sensing stage. Red refers to the actual logic that experiences N/PBTI. Odd number of stages can form a ring oscillator and serve as test structures or sensors. (**b**) Truth table for the sensing stage. When under NBTI or PBTI mode, PMOS or NMOS transistors are under constant stress; when under test mode, the structure works as a regular ring oscillator stage, and the oscillation frequency can be read out from the RO output

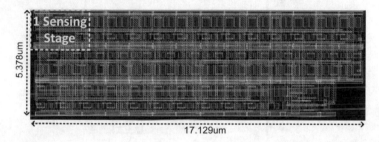

Fig. 4.24 Layout of an example of using sensing element in a 37-stage ring oscillator in 28 nm FD-SOI technology. Overall area is small, and the sensor can be configured to sense either NBTI or PBTI. It can also be used as the test structure for studying N/PBTI degradations

than the same RO configuration that is implemented with single inverter only. The structure can be used as a building block for either NBTI or PBTI sensing scheme, and it can also be employed as a BTI test structure for characterizing N/PBTI and understanding the differences between the two mechanisms.

4.4.3 Metastable-Element-Based Embeddable BTI Sensors

The critical path determines the performance of the system, but, due to the fact that BTI wearout is a gradual effect that accumulates with time, different paths experience various switching activities and/or temperatures. It is thus highly possible that some non-critical paths that are only marginally faster than the critical path become the new critical path (and vice versa) in multi-output cases [32]. Figure 4.25 gives an example of such scenario. The circuit shown in Fig. 4.25a

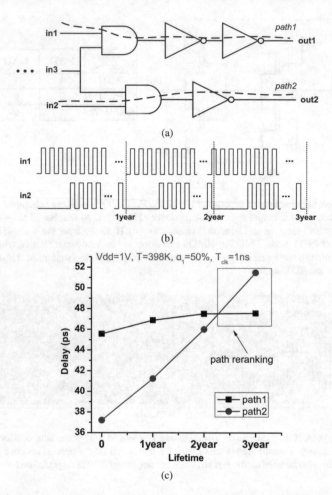

Fig. 4.25 Illustration of the BTI-induced critical path reranking. (**a**) Simulated circuit. (**b**) Input patterns for the circuit. (**c**) Path delay after BTI stress. Path reranking happens after around 2.3 years, the delay of path 2 surpasses path 1

is simulated in a 28 nm FD-SOI technology node. BTI-related wearout parameters for that technology are extracted based on the experimental results from [48, 49]. Assume $path1$ is (part of) the critical path, and $path2$ is the second slowest path in a circuit. Given different input patterns for the two path as shown in Fig. 4.25b, simulated results in Fig. 4.25c show that after certain times, the delay of $path2$ surpasses $path1$, and this results in so-called path reranking (or reordering). The conventional ways of dealing with BTI wearout and sensing BTI mostly focused on the critical path without considering the path reranking issues. If this is left unchecked, the system may experience an unexpected failure. For BTI sensors, tracking more paths, especially the second or third most critical path is thus necessary to guarantee the expected performance at the system level. The most obvious way to do this is to use individual sensors for each path, but due to variations (like process) of these sensors, one path usually needs multiple sensors, the area overhead becoming unacceptable; thus compact sensors that are able to track multiple paths simultaneously are preferred in this case. This section presents such a sensor which is able to track multiple paths simultaneously within one embeddable structure. The sensor is able to track BTI recovery so that any proactive recovery techniques discussed in the previous chapters can be enabled by the sensor outputs. The proposed sensors are implemented and simulated in the same industry standard 28 nm FD-SOI technology. Details will be presented in the following sections.

4.4.3.1 Sensor Scheme

Figure 4.26 shows the main topology of the proposed NBTI sensors. The sensor is based on the metastable cell which composes of a degradation inverter and a reference inverter [41, 42]. Driven by the paths of interests, the sensor is exposed to the same environment as the core circuit, and degrades at the same rate as the paths. Figure 4.26a shows the high-level block drives several cells for active recovery purpose. Figure 4.26b is the case when multiple paths drive separate cells, and is aware of path reranking. In both cases, only one reference inverter is needed and can be shared by all tracking inverters; the outputs can also be used as trigger signals for DVFS or body biasing techniques to mitigate the effect of wearout. Figure 4.27 shows the transistor-level implementation of the proposed sensor which merges Fig. 4.26a, b into a single design. The bottom half of the schematic corresponds to Fig. 4.26a, where $P0$–$P2$ are tracking transistors that are sized based on the degradation percentage (2, 5, and 10% in this case). The percentage is determined by

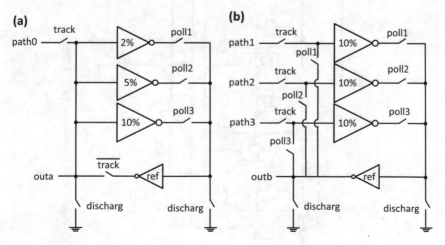

Fig. 4.26 Multiple-critical-path embeddable BTI sensor high-level scheme. (**a**) Proactive recovery sensing; (**b**) multiple paths sensing

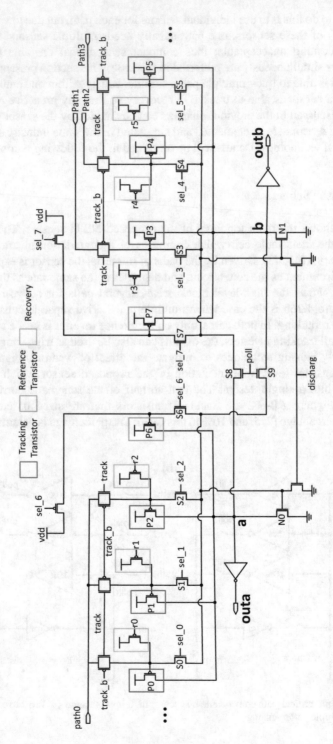

Fig. 4.27 Transistor-level schematic of the BTI sensor

the driven strength of the transistor, which further corresponds to the size. Similarly, the top half corresponds to Fig. 4.26b which enables multiple paths tracking with the same degradation percentage limit (5%) for detecting the path reranking issues. Each half has a reference transistor ($P6$, $P7$). The binary outputs a and b are read out through inverters. The transmission gates between tracking transistors and paths are used for selecting certain path, and also reducing the critical path performance impact due to the sensor. The symmetric structure is able to leverage the load imbalance between nodes a and b. In this implementation, the sensor is for NBTI tracking; PBTI tracking can be enabled by replacing all the PMOS logic to the corresponding NMOS logic. The design shown in the figure can track four paths. More can be extended by copying one branch and adding to the existing one as needed.

Figure 4.28 shows the sensor functionality through simulation. The inputs can be encoded to only six signals. The sensor has three modes, tracking mode, polling mode, and recovery mode. In tracking mode, the track signal is set high and tracking transistors are stressed by different paths. The reference transistors are OFF to stay "fresh." NMOS switches $S0$–$S7$ are turned OFF to separate the tracking transistors from the polling circuitry. In polling mode, the *track* signal is set low, $S6$ is turned ON if the right-half paths need to be polled; similarly, $S7$ corresponds to the left-half paths. $S0$–$S5$ are turned ON one by one, during each ON period (polling period), both outputs (a and b) are discharged quickly, and then the poll signal is set high, a fight condition is created between the tracking transistors and the reference transistors; since the tracking transistors are sized slightly larger (stronger) initially, assume that the loads seen from each inverter are equal, the stronger ones will "win the fights" and latch the outputs to high. When NBTI-induced degradation reaches a point where reference transistors win the fights, it means the expected degradation

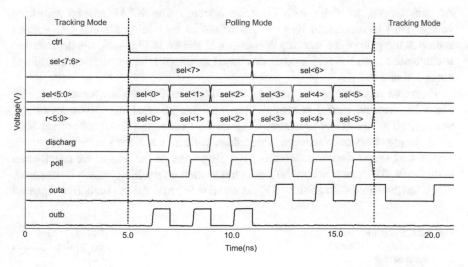

Fig. 4.28 Functionality simulation of the BTI sensor under fresh ($time = 0$) condition

limit is detected and the sensor is triggered. The total polling time for all paths is ~ 12 ns, and it is clear that it cannot introduce significant NBTI-induced degradation or recovery from a long-term perspective. Recovery mode works similar to the tracking mode, $r0$–$r5$ are recovery control signals for each path. When recovery is scheduled, these signals are set to low, and the tracking transistors are in recovery mode; during these recovery periods, polling is still needed to check if the expected recovery level is achieved. A nice feature of the sensor is that any recovery boosting techniques discussed in Chap. 2 can also easily be applied to this sensor so that it is under the exact same recovery conditions as the active circuitry.

4.4.3.2 Sizing

Size corresponds to the strength of the transistors. Tracking transistors are sized $x\%$ larger (x is set as the threshold for BTI wearout) than reference transistors using multi-fingers for easy area changes. Since variations will affect the accuracy of the sensors, we pick the finger size (120 nm) a bit larger than the minimum (80 nm in 28 nm FD-SOI technology). All switches are sized equally. Two footer NMOS transistors ($N0$ and $N1$) in the metastable cell are also sized equally and relatively large to have a fast discharging. The load imbalance is dealt with by adding dummy transistors to the load as suggested in [41]. Variations are also dealt with by using interdigitation techniques for degradation and reference transistors in the physical design phase.

4.4.3.3 Proactive Recovery Case Simulation Results

All simulations are done with Cadence Spectre. The NBTI-induced threshold voltage shift is introduced into the circuit netlist by adding a small DC voltage source at the gate of the tracking transistors as shown in Fig. 4.29; the degradation is calculated from the BTI models discussed in Chap. 2. The operating conditions are at 1 V and 398 K.

Proactive recovery corresponds to the bottom half of the sensor as shown in Fig. 4.27. Assume $path0$ is always "low" so that all tracking transistors on the bottom half are under constant stress. Figure 4.30 shows the threshold voltage shift ΔV_{th} during stress and recovery, where flags indicate when the sensor is triggered. Figure 4.31 shows the corresponding triggering order of the sensor; we check $outa$ in this case. ① corresponds to the fresh status when no tracking branch is triggered. When the threshold voltage shift ΔV_{th} is equal to 9.5 mV, the 2% branch is triggered

Fig. 4.29 Simulation setup for introducing BTI wearout in a circuit netlist

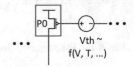

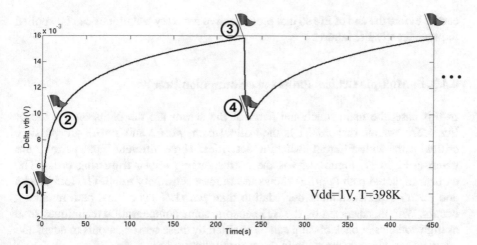

Fig. 4.30 Threshold degradation and sensor trigger point

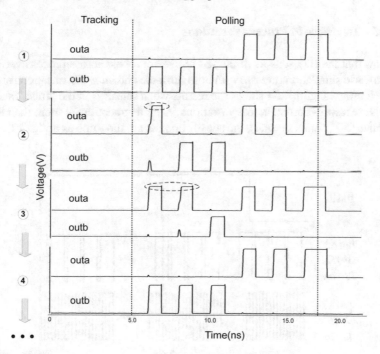

Fig. 4.31 Sensor triggering order for the proactive recovery case

(②), and ΔV_{th} keeps increasing until 5% threshold is hit (③, ΔV_{th} is 15.8 mV). Then the recovery signal is triggered, the recovery starts until it reaches the point when all branches are not triggered (④), and the process repeats. Instead of waiting until 10% path is triggered (ΔV_{th} is 25.4 mV), the proposed sensor is triggered much

earlier before the end of life so that proper active recovery techniques can be applied
to prevent early-life failures.

4.4.3.4 Multiple Critical Paths Case Simulation Results

In this case, the main functional parts of the sensor are the paths on the top of
Fig. 4.27. Assume that *path*1 is the critical path, *path*2 and *path*3 are possible
critical paths with different switching activities. Three different input patterns are
given in Fig. 4.32. Figure 4.33 is the corresponding sensor triggering order. The
earliest triggered path (*path*2) shows the highest sensitivity to NBTI. Both *path*2
and *path*3 experience more degradation than *path*1. Thus critical path reranking
occurs. With this detected by the BTI sensors, some compensation techniques such
as high voltages or back biasing can be enabled by these sensor outputs to adapt and
avoid performance degradation in the early lifetime.

4.4.3.5 Tolerance to Process Variations

To show that the BTI sensors discussed above are robust across process variations,
Monte Carlo simulations are run with both intra-die and inter-die variations modeled
by the foundry. Figure 4.34 shows the results when sensor is in the "fresh" state; the
outputs are always correct across variations. After degradation, at more than 85% of
the points (500 points in total), the sensor outputs are the same as at the TT corner.

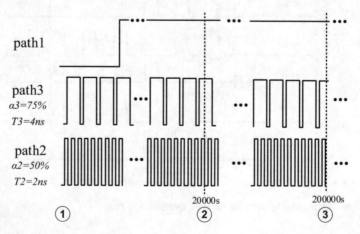

Fig. 4.32 Input patterns for multiple path simulation case

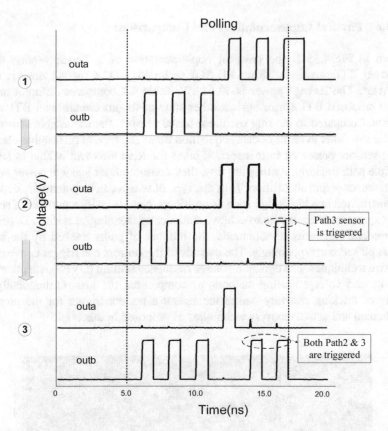

Fig. 4.33 Sensor trigger order in multiple critical paths case

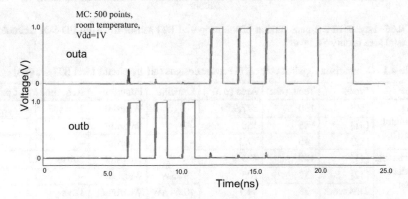

Fig. 4.34 Monte Carlo simulations at the "fresh" status (500 points). More than 85% of the seeds give the correct outputs. The overlapping color in the figure refers to waveforms for each seed. The output of the sensor at TT corner is used as the reference

4.4.3.6   Physical Implementations and Comparisons

Shown in Fig. 4.35 is the physical implementation of a 2-path version of the proposed BTI sensor in 28 nm FD-SOI technology. The overall area is about $54.1 \, \mu m^2$. The leakage power is 40.64 nW. Table 4.1 compares different metrics of the proposed BTI sensor against other state-of-the-art circuit-level BTI sensor designs. Compared to the ring oscillator-based sensors, the metastable sensors are smaller and faster in terms of data acquisition time. The type of metastable-element-based sensors presented in this section takes the least area and is able to support multiple path tracking. As they are tiny, they consume least leakage power and are fast in sensor output acquisition. Thus this type of sensor can be potentially deployed and distributed in a big design such as multicore systems, with a number of sensors being spread over each core to achieve statistically meaningful results. Depending on area and reliability requirements, the number of paths tracked by the sensor can be picked correspondingly. The outputs of the sensors can trigger conventional adaptive techniques, like dynamic voltage frequency scaling (DVFS), adaptive body biasing, and voltage scaling methods to compensate the loss. Orthogonally, the ability of tracking recovery makes the sensor a perfect trigger for the proactive accelerated and active recovery techniques as proposed in this book.

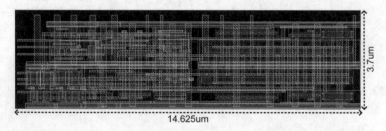

14.625um

Fig. 4.35 Layout of a 2-path version of the proposed BTI sensor in 28 nm FD-SOI technology. The total area is only $54.1 \, \mu m^2$

Table 4.1 Comparisons against other BTI sensor designs (all are circuit-level BTI sensors)

Type	Work	Tech. (nm)	Area (μm^2)	Leakage	Function	Acq. time	# of paths
Ring oscillator based	[50]	130	277,950	–	Wearout	29 μs	–
	[51]	45	150	–	Wearout	–	1
	[52]	45	77.3	–	Wearout	–	1
Metastable element based	[41]	150	493.2	–	Wearout	200 ns	1
	[42]	32	105	239 nW	Wearout	–	1
	This work	28	54.1	40.64 nW	Wearout/recovery	12 ns	2

4.4.3.7 Sensor Placement Methodology

Including the proposed sensor into a top-down ASIC design flow will be more efficient than the custom design solution, and it also enables design reuse, especially when a large number of sensors are distributed in a complex circuit. All instances (single transistors, transmission gates, or inverters) of the sensor can be used directly from the standard cell library of a process design kit (PDK)—this makes it easy to implement the sensors directly with the conventional PnR tools. Figure 4.36 shows an example design which is directly placed and routed with the Synopsys IC Compiler in a 28/30 nm technology node. The design is relative bigger compared to the custom designed one shown in Fig. 4.35, but this is expected. In this case, the area can be further reduced by decreasing the utilization percentage and putting more constraints for the tool. The sensor can be used as a design IP that is no different from other cells in the standard cell library.

As many such BTI sensor IPs are expected to be placed in a design, a methodology that is able to automatically distribute the sensor is necessary. In this book, we present one such candidate. The compact sensor can be embedded into a scan chain cell that is normally used for design for test (DFT) purposes. Figure 4.37 illustrates the idea, where the newly designed scan cell has the sensor embedded. Figure 4.38 is the design methodology of integrating the sensor with the scan cell from the standard cell library in a Synopsys design environment. Design Compiler (DC) is used for synthesis, the output being the synthesized sensor netlist which can be fed into IC Compiler (ICC) for PnR and generating the sensor IP layout (in .CEL and .FRAM format for Synopsys tools). The new *scan cell* netlist is edited by adding the *scan cell* component to the synthesized sensor netlist. Then the new scan cell can be placed and routed. With this methodology, a modified version of scan cell can be treated as a building block for a scan chain.

As shown in Fig. 4.37, the sensor readout can be selected by a MUX logic. It can be used in two ways. In the closed-loop case, run-time solutions like adaptive or accelerated active recovery can be enabled based on the sensor outputs; in the open-loop case, sensor outputs can be scanned out to the system just as how regular scan chains work so that the system can conduct statistical analysis or just be aware of the wearout and recovery behaviors of internal nodes, and corresponding decisions

Fig. 4.36 A 2-path version of the sensor implemented with the PnR tool directly

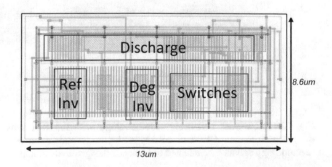

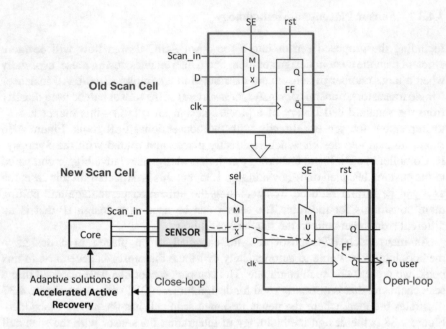

Fig. 4.37 Embedded sensor in a scan chain cell in support of both closed-loop and open-loop sensor readouts

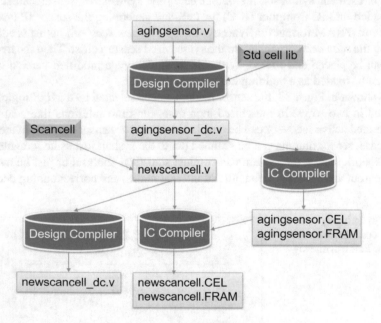

Fig. 4.38 New scan cell design methodology with the BTI sensor embedded in a Synopsys design environment. .CEL and .FRAM are layout formats required by the Synopsys tools. Similar methodology can also be adapted to Cadence design environment

can be made off-line based on the sensor outputs. In Appendix A, we present the detailed flow and an example of instrumenting the BTI sensors in a counter design.

4.5 EM Sensing

Compared to BTI sensing, EM sensing is more challenging because of several reasons. First, EM behavior is more complex as shown in Chap. 3; due to the stress accumulation period, the resistance stays flat for a long time, with relative sudden change after stress accumulates. This indicates that EM sensors might need to be ON more frequently than BTI sensors so that any EM-triggered event can be recorded precisely and timely. Second, EM mostly occurs in power delivery network, and it is far from the front-end-of-line, which is the logic. Thus any logic-style sensor will not be able to experience exactly the same conditions as the metals on the very top. Finally, differentiating EM wearout from other wearout phenomena such as BTI is challenging since they both result in similar degradation effects (slow down) in many cases. In this section, we mainly review some of the existing EM sensor designs and discuss how these sensors can be applied to the accelerated and active EM recovery purposes.

Since EM increases IR drop, it can lead to load performance degradations, thus a simple way of detecting EM is to use a ring oscillator as the load and check the frequency degradation [53]. Since this method shares many similarities to existing BTI sensing techniques as detailed in Sect. 4.4.1, it is sometime challenging to separate EM from other wearout-induced degradations. Ring oscillators can be perfect sensor candidates in the case when one cares about the impact of wearout as a whole, and not about separating the impact of each wearout mechanism. Metal-line-based sensor appears to be a better candidate for EM-specific sensing and detection [54, 55]. The main idea is to put a set of metal lines on chip so that they can experience the same or more EM than the power delivery networks or other EM-sensitive interconnects. Figure 4.39 gives an example of one such EM sensing structure that is modified based on the previous work [54, 55]; they are essentially a set of on-chip metal lines in parallel. Since there are inherent variations in the metal wires, multiple metals are required to get statistically meaningful results. A metal line that is kept unstressed is used as the reference metal for capturing any resistance change. The idea behind this is somewhat similar to the metastable-element-based BTI sensor discussed in the previous section. The dimensions of these metal lines need to be picked based on the system lifetime requirement, which can be translated into a resistance increase threshold such as 10%. Thus the current densities in the sensing metal lines have to be relatively higher than in regular loads in order to force a shorter lifetime, and also in order to compensate for the reduced number of sensing elements compared to regular loads. Intermediate sensing check points (e.g., 5% of resistance increase) can be added by designing metal structures with different width values as shown in Fig. 4.39. The current that flows through the sensing wires, although scaled up, has to be proportional with the current in the

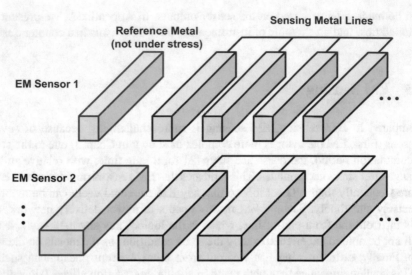

Fig. 4.39 Illustration of metal-line-based EM sensors. Multiple dimensions can be used to sense at different levels

Fig. 4.40 The EM-induced resistance change detection circuit (design is modified based on the circuit proposed in [55])

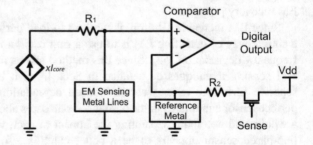

main circuit in order to have a high correlation. Finally local operating conditions such as temperatures need to be reflected in the EM sensors, so multiple copies of such sensors need to be placed accordingly.

In order to sense the EM degradations during run time, a resistance detection circuitry is necessary. He et al. [55] proposed one such scheme, a modified version is presented in Fig. 4.40. The current source on the left is mirrored from the current of the actual loads—it provides the stress to the sensing metal lines shown in Fig. 4.39. A comparator is used for comparing the voltage levels across the sensing structures and the reference metal line. If the voltage across the sensing structure is higher, then it means the EM-induced resistance increase has reached the threshold, and the sensor is triggered. The *sense* signal stays high during the sensing periods and the reference metal line is unstressed; it is set low only during the sensor readout periods. As can be seen from Fig. 4.40, introducing analog circuits such as comparators and on-chip resistors can lead to more design effort compared to the BTI sensor design. Also EM sensing circuitry will take more silicon area. The good

Fig. 4.41 Illustrations of EM sensor usage in the accelerated and active recovery case

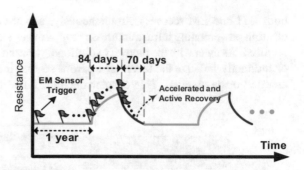

news is that as EM is usually a concern mostly for power and ground network, the number of sensors required is much less than in the BTI case where each transistor is undergoing wearout. Also the sensing structures itself are only a set of metal lines that take relatively small area. As reported in [56], one such EM sensor with 10 stressed wires occupies only 100–500 μm^2 in a 180 nm technology. In addition, the metal line sensing structures can be distributed in a fine-granularity way, and the resistance detection circuitry can be shared by multiple copies of such sensing structures with the extra selection logic.

As suggested by the experimental results discussed in Chap. 3, EM has a long period of stress accumulation (e.g., 1 year) when the resistance stays almost unchanged; thus EM sensors can be in *sense* mode most of the time and the output of the sensor can be sampled every few days or longer. In the second region (void nucleation) of EM, when the resistance starts to increase relatively abruptly, the sensor needs to be alerted more frequently to sense the converting point between the first and second region, and also track the resistance changes. Accelerated and active EM recovery techniques can be applied after resistance increase to a threshold that was set based on analysis. During recovery periods, EM sensors need to be checked frequently as well to avoid the potential EM issues by reverse current (as shown in Fig. 3.8 in Chap. 3). This has been illustrated in Fig. 4.41. In summary, metal-line-based EM sensors can serve as a check engine and indicator for EM wearout and recovery, they can be integrated as a circuit element IP for an accelerated self-healing system.

4.6 Conclusions

In this chapter, we presented a set of circuit structures for enabling and assisting the accelerated and active BTI/EM recovery. The key components are highlighted in Fig. 4.42. On-chip negative voltage is designed for generating the negative voltage for activating the BTI recovery; on-chip tunable heater is a modified version of configurable ring oscillator which can be deployed for providing high temperature when necessary. A multi-mode EM/BTI recovery assist circuit is able to support both BTI and EM recovery simultaneously, and it can be designed based on the existing power gating infrastructure on chip, thus less design effort and overhead are required. As even with the recovery solutions, wearout sensors are needed to directly or indirectly indicate the levels of wearout so that these recovery solutions can be asserted or de-asserted. This chapter detailed three different sensor designs—two

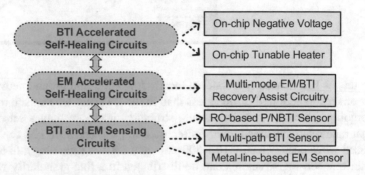

Fig. 4.42 Chapter 4 highlights

Table 4.2 Summary of PPA metrics for different circuit components (in 28 nm FD-SOI technology unless specified)

Type	Design name	Leakage power	Dynamic power	Area	Performance
BTI accelerated and active recovery circuits	Neg. voltage generator[a]	68.85 nW	64.47 μW	4300 μm^2	>66.7 MHz
	On-chip heater[b]	16.8 nW	75 μW	16 μm^2	–
EM accelerated and active recovery circuits	Multi-mode assist circuit[c]	–	–	58.24 μm^2	Wake-up time ~ 170 ns
BTI sensors	RO-based P/NBTI sensors	19 nW	–	92 μm^2	–
	Metastable element-based sensors[d]	40.64 nW	–	54.1 μm^2	Acq. time 12 ns
EM sensors	Metal-line-based sensors[e]	–	–	100–500 μm^2	–

[a]Corresponds to a negative voltage generator designed for generating −0.3 V for BTI active recovery

[b]Corresponds to one on-chip heater (41-stage RO) that can generate temperature of more than 80 °C, in the real use case, multiple copies of this heater will be distributed

[c]PPA metrics of this circuit depends on the load size and application, here we list the examples for 8 ring oscillators running in parallel

[d]The numbers presented in this table show metrics for a 2-path version of the sensor

[e]Data is reported by He [56] with 180 nm technology node

for BTI and one for EM. These sensors are small and flexible, and can be adapted to multiple use scenarios. In the case of proactive recovery, sensors are for sanity check so that some wearout effects (e.g., EM) are not "overly-recovered"; in the reactive recovery cases, all the actuations rely on these sensor outputs. As BTI and EM wearout effects become increasingly critical, novel techniques that are able to mitigate them with lower overhead are highly desirable. Table 4.2 summarizes the power, performance, and area metrics for the circuit components discussed in this chapter. It shows that most of these components are small in terms of area and fast in terms of response time. In summary, this chapter provides a set of circuit IP blocks and infrastructures for designing future accelerated self-healing systems. With these components at the circuit level, smart design decisions at higher levels of a system stack can be made to leverage the trade-offs to support self-healing and reliable system design. We will further discuss several of these candidate solutions in the next chapter.

References

1. Kaushik Roy, Saibal Mukhopadhyay, and Hamid Mahmoodi-Meimand. Leakage current mechanisms and leakage reduction techniques in deep-submicrometer cmos circuits. *Proceedings of the IEEE*, 91(2):305–327, 2003.
2. Piotr Weber, Maciej Zagrabski, Przemyslaw Musz, Krzysztof Kepa, Maciej Nikodem, and Bartosz Wojciechowski. Configurable heat generators for fpgas. In *Thermal Investigations of ICs and Systems (THERMINIC), 2014 20th International Workshop on*, pages 1–4. IEEE, 2014.
3. S Sarma, N Dutt, N Venkatasubramanian, A Nicolau, and P Gupta. Cyberphysical system-on-chip (cpsoc): Sensor actuator rich self-aware computational platform. *University of California Irvine, Tech. Rep. CECS TR-13-06*, 2013.
4. Maxim Switched-Capacitor Voltage Converters MAX1044/ICL7660 datasheet:. https://www.maximintegrated.com/en/products/power/charge-pumps/ICL7660.html.
5. Ajith Sivadasan, Florian Cacho, Sidi Ahmed Benhassain, Vincent Huard, and Lorena Anghel. Study of workload impact on bti hci induced aging of digital circuits. In *Proceedings of the 2016 Conference on Design, Automation & Test in Europe*, pages 1020–1021. EDA Consortium, 2016.
6. Andrea Calimera, Alberto Macii, Enrico Macii, and Massimo Poncino. Power-gating for leakage control and beyond. In *Circuit Design for Reliability*, pages 175–205. Springer, 2015.
7. Andrea Calimera, Enrico Macii, and Massimo Poncino. Nbti-aware sleep transistor design for reliable power-gating. In *Proceedings of the 19th ACM Great Lakes symposium on VLSI*, pages 333–338. ACM, 2009.
8. Andrea Calimera, Enrico Macii, and Massimo Poncino. Design techniques for nbti-tolerant power-gating architectures. *Circuits and Systems II: Express Briefs, IEEE Transactions on*, 59(4):249–253, 2012.
9. Kai Ma and Xiaorui Wang. Pgcapping: exploiting power gating for power capping and core lifetime balancing in cmps. In *Proceedings of the 21st international conference on Parallel architectures and compilation techniques*, pages 13–22. ACM, 2012.
10. Fabian Oboril and Mehdi B Tahoori. Extratime: A framework for exploration of clock and power gating for bti and hci aging mitigation. *ITG-Fachbericht-Zuverlässigkeit und Entwurf*, 2011.

11. Tuck-Boon Chan, John Sartori, Puneet Gupta, and Rakesh Kumar. On the efficacy of nbti mitigation techniques. In *Design, Automation & Test in Europe Conference & Exhibition (DATE), 2011*, pages 1–6. IEEE, 2011.
12. Ming-Chao Lee, Yu-Guang Chen, Ding-Kei Huang, and Shih-Chieh Chang. Nbti-aware power gating design. In *Design Automation Conference (ASP-DAC), 2011 16th Asia and South Pacific*, pages 609–614. IEEE, 2011.
13. Kai-Chiang Wu, Diana Marculescu, Ming-Chao Lee, and Shih-Chieh Chang. Analysis and mitigation of nbti-induced performance degradation for power gated circuits. In *Proceedings of the 17th IEEE/ACM international symposium on Low-power electronics and design*, pages 139–144. IEEE Press, 2011.
14. Kai-Chiang Wu, Chao Lin, Yao-Te Wang, and Shuen-Shiang Yang. Bti-aware sleep transistor sizing algorithm for reliable power gating designs. *Computer-Aided Design of Integrated Circuits and Systems, IEEE Transactions on*, 33(10):1591–1595, 2014.
15. Daniele Rossi, Vasileios Tenentes, Bashir Al-Hashimi, et al. Nbti and leakage aware sleep transistor design for reliable and energy efficient power gating. *Proceedings of the IEEE European Test Symposium*, 2015.
16. Norihiro Kamae, Akira Tsuchiya, and Hidetoshi Onodera. A body bias generator compatible with cell-based design flow for within-die variability compensation. In *Solid State Circuits Conference (A-SSCC), 2012 IEEE Asian*, pages 389–392. IEEE, 2012.
17. Norihiro Kamae, AKM Mahfuzul Islam, Akira Tsuchiya, and Hidetoshi Onodera. A body bias generator with wide supply-range down to threshold voltage for within-die variability compensation. In *Solid-State Circuits Conference (A-SSCC), 2014 IEEE Asian*, pages 53–56. IEEE, 2014.
18. Xrysovalantis Kavousianos, Krishnendu Chakrabarty, Arvind Jain, and Rubin Parekhji. Test schedule optimization for multicore socs: Handling dynamic voltage scaling and multiple voltage islands. *IEEE Transactions on Computer-Aided Design of Integrated Circuits and Systems*, 31(11):1754–1766, 2012.
19. A. Amouri, J. Hepp, and M. Tahoori. Built-in self-heating thermal testing of fpgas. *IEEE Transactions on Computer-Aided Design of Integrated Circuits and Systems*, PP(99):1–1, 2016.
20. Abdulazim Amouri, Jochen Hepp, and Mehdi Tahoori. Self-heating thermal-aware testing of fpgas. In *VLSI Test Symposium (VTS), 2014 IEEE 32nd*, pages 1–6. IEEE, 2014.
21. Kaiyuan Yang, Qing Dong, Wanyeong Jung, Yiqun Zhang, Myungjoon Choi, David Blaauw, and Dennis Sylvester. 9.2 a 0.6 nj- 0.22/+ 0.19° c inaccuracy temperature sensor using exponential subthreshold oscillation dependence. In *Solid-State Circuits Conference (ISSCC), 2017 IEEE International*, pages 160–161. IEEE, 2017.
22. Xiaoyang Wang, Po-Han Peter Wang, Yuan Cao, and Patrick P Mercier. A 0.6 v 75nw all-cmos temperature sensor with 1.67 m° c/mv supply sensitivity. *IEEE Transactions on Circuits and Systems I: Regular Papers*, 2017.
23. Xin Huang, Valeriy Sukharev, Taeyoung Kim, and Sheldon X-D Tan. Dynamic electromigration modeling for transient stress evolution and recovery under time-dependent current and temperature stressing. *Integration, the VLSI Journal*, 2016.
24. V Sukharev, X Huang, and SX-D Tan. Electromigration induced stress evolution under alternate current and pulse current loads. *Journal of Applied Physics*, 118(3):034504, 2015.
25. Jaume Abella, Xavier Vera, Osman S Unsal, Oguz Ergin, Antonio González, and James W Tschanz. Refueling: Preventing wire degradation due to electromigration. *IEEE micro*, (6):37–46, 2008.
26. Aditya Bansal and Jae-Joon Kim. Power napping technique for accelerated negative bias temperature instability (nbti) and/or positive bias temperature instability (pbti) recovery, July 21 2015. US Patent 9086865.
27. Deepak C Sekar, Bing Dang, Jeffrey A Davis, and James D Meindl. Electromigration resistant power delivery systems. *IEEE electron device letters*, 28(8):767–769, 2007.
28. Sanjay V Kumar, Chris H Kim, and Sachin S Sapatnekar. Adaptive techniques for overcoming performance degradation due to aging in digital circuits. In *Proceedings of the Asia and South Pacific Design Automation Conference*, pages 284–289. IEEE Press, 2009.

29. Hassan Mostafa, Mohab Anis, and Mohamed Elmasry. Nbti and process variations compensation circuits using adaptive body bias. *Semiconductor Manufacturing, IEEE Transactions on*, 25(3):460–467, 2012.
30. Eric Karl, Dennis Sylvester, and David Blaauw. Analysis of system-level reliability factors and implications on real-time monitoring methods for oxide breakdown device failures. In *Quality Electronic Design, 2008. ISQED 2008. 9th International Symposium on*, pages 391–395. IEEE, 2008.
31. Evelyn Mintarno, Vishal Chandra, David Pietromonaco, Robert Aitken, and Robert W Dutton. Workload dependent nbti and pbti analysis for a sub-45nm commercial microprocessor. In *Reliability Physics Symposium (IRPS), 2013 IEEE International*, pages 3A–1. IEEE, 2013.
32. Wenping Wang, Shengqi Yang, Sarvesh Bhardwaj, Rakesh Vattikonda, Sarma Vrudhula, Frank Liu, and Yu Cao. The impact of nbti on the performance of combinational and sequential circuits. In *Proceedings of the 44th annual Design Automation Conference*, pages 364–369. ACM, 2007.
33. Navid Khoshavi, Rizwan A Ashraf, Ronald F DeMara, Saman Kiamehr, Fabian Oboril, and Mehdi B Tahoori. Contemporary CMOS aging mitigation techniques: Survey, taxonomy, and methods. *Integration, the VLSI Journal*, 59:10–22, 2017.
34. Mitsuhiko Igarashi, Kan Takeuchi, Takeshi Okagaki, Koji Shibutani, Hiroaki Matsushita, and Koji Nii. An on-die digital aging monitor against hci and xbti in 16 nm fin-fet bulk cmos technology. In *European Solid-State Circuits Conference (ESSCIRC), ESSCIRC 2015-41st*, pages 112–115. IEEE, 2015.
35. Tae-Hyoung Kim, Randy Persaud, and Chris H Kim. Silicon odometer: An on-chip reliability monitor for measuring frequency degradation of digital circuits. *Solid-State Circuits, IEEE Journal of*, 43(4):874–880, 2008.
36. Dipak Sengupta and Sachin S Sapatnekar. Predicting circuit aging using ring oscillators. In *Design Automation Conference (ASP-DAC), 2014 19th Asia and South Pacific*, pages 430–435. IEEE, 2014.
37. Woongrae Kim, Taizhi Liu, and Linda Milor. On-line monitoring of system health using on-chip srams as a wearout sensor. In *On-Line Testing and Robust System Design (IOLTS), 2017 IEEE 23rd International Symposium on*, pages 253–258. IEEE, 2017.
38. M Denais, C Parthasarathy, G Ribes, Y Rey-Tauriac, N Revil, A Bravaix, V Huard, and F Perrier. On-the-fly characterization of nbti in ultra-thin gate oxide pmosfet's. In *Electron Devices Meeting, 2004. IEDM Technical Digest. IEEE International*, pages 109–112. IEEE, 2004.
39. Kunhyuk Kang, Keejong Kim, Ahmad E Islam, Muhammad Alam, Kaushik Roy, et al. Characterization and estimation of circuit reliability degradation under nbti using on-line iddq measurement. In *Design Automation Conference, 2007. DAC'07. 44th ACM/IEEE*, pages 358–363. IEEE, 2007.
40. Jiangyi Li and Mingoo Seok. Robust and in-situ self-testing technique for monitoring device aging effects in pipeline circuits. In *Proceedings of the 51st Annual Design Automation Conference*, pages 1–6. ACM, 2014.
41. Stuart N Wooters, Adam C Cabe, Zhenyu Qi, Jiajing Wang, Randy W Mann, Benton H Calhoun, Mircea R Stan, and Travis N Blalock. Tracking on-chip age using distributed, embedded sensors. *Very Large Scale Integration (VLSI) Systems, IEEE Transactions on*, 20(11):1974–1985, 2012.
42. Vikram B Suresh and Wayne P Burleson. Fine grained wearout sensing using metastability resolution time. In *Quality Electronic Design (ISQED), 2014 15th International Symposium on*, pages 480–485. IEEE, 2014.
43. Hossein Karimiyan Alidash, Andrea Calimera, Alberto Macii, Enrico Macii, and Massimo Poncino. On-chip nbti and pbti tracking through an all-digital aging monitor architecture. In *PATMOS*, pages 155–165. Springer, 2012.

44. Abhishek Koneru, Arunkumar Vijayan, Krishnendu Chakrabarty, and Mehdi B Tahoori. Fine-grained aging prediction based on the monitoring of run-time stress using dft infrastructure. In *Computer-Aided Design (ICCAD), 2015 IEEE/ACM International Conference on*, pages 51–58. IEEE, 2015.
45. Farshad Firouzi, Fangming Ye, Krishnendu Chakrabarty, and Mehdi B Tahoori. Aging-and variation-aware delay monitoring using representative critical path selection. *ACM Transactions on Design Automation of Electronic Systems (TODAES)*, 20(3):39, 2015.
46. S Zafar, YH Kim, V Narayanan, C Cabral Jr, V Paruchuri, B Doris, J Stathis, A Callegari, and M Chudzik. A comparative study of nbti and pbti (charge trapping) in sio2/hfo2 stacks with fusi, tin, re gates. In *VLSI Technology, 2006. Digest of Technical Papers. 2006 Symposium on*, pages 23–25. IEEE, 2006.
47. Tony Tae-Hyoung Kim, Pong-Fei Lu, Keith A Jenkins, and Chris H Kim. A ring-oscillator-based reliability monitor for isolated measurement of nbti and pbti in high-k/metal gate technology. *IEEE Transactions on Very Large Scale Integration (VLSI) Systems*, 23(7):1360–1364, 2015.
48. J El Husseini, A Subirats, X Garros, A Makoseij, O Thomas, G Reimbold, V Huard, F Cacho, and X Federspiel. Accurate modeling of dynamic variability of sram cell in 28 nm fdsoi technology. In *Microelectronic Test Structures (ICMTS), 2014 International Conference on*, pages 41–46. IEEE, 2014.
49. D Angot, V Huard, X Federspiel, F Cacho, and A Bravaix. Bias temperature instability and hot carrier circuit ageing simulations specificities in UTBB FDSOI 28nm node. In *Reliability Physics Symposium (IRPS), 2013 IEEE International*, pages 5D–2. IEEE, 2013.
50. John Keane, Tae-Hyoung Kim, and Chris H Kim. An on-chip nbti sensor for measuring pmos threshold voltage degradation. *IEEE transactions on very large scale integration (VLSI) systems*, 18(6):947–956, 2010.
51. John Keane, Xiaofei Wang, Devin Persaud, and Chris H Kim. An all-in-one silicon odometer for separately monitoring hci, bti, and tddb. *IEEE Journal of Solid-State Circuits*, 45(4):817–829, 2010.
52. Prashant Singh, Eric Karl, David Blaauw, and Dennis Sylvester. Compact degradation sensors for monitoring nbti and oxide degradation. *IEEE Transactions on Very Large Scale Integration (VLSI) Systems*, 20(9):1645–1655, 2012.
53. Xuehui Zhang, Kan Xiao, and Mohammad Tehranipoor. Path-delay fingerprinting for identification of recovered ics. In *Defect and Fault Tolerance in VLSI and Nanotechnology Systems (DFT), 2012 IEEE International Symposium on*, pages 13–18. IEEE, 2012.
54. Mircea R Stan and Paolo Re. Electromigration-aware design. In *Circuit Theory and Design, 2009. ECCTD 2009. European Conference on*, pages 786–789. IEEE, 2009.
55. Kai He, Xin Huang, and Sheldon X-D Tan. Em-based on-chip aging sensor for detection and prevention of counterfeit and recycled ics. In *Computer-Aided Design (ICCAD), 2015 IEEE/ACM International Conference on*, pages 146–151. IEEE, 2015.
56. Kai He. *Parallel CAD Algorithms and Hardware Security for VLSI Systems*. PhD thesis, University of California, Riverside, 2016.

Chapter 5
Active Accelerated Self-healing as a Key Design Knob for Cross-Layer Resilience

5.1 Overview

In the previous chapter we presented several circuit-level techniques for activating and accelerating BTI and EM recovery. As many of these single-layer solutions (e.g., circuit only) may not be economic in terms of the introduced overhead, and also need to be triggered in a smart way, dealing with wearout issues and scheduling the accelerated and active recovery periods need to be done across layers, where various techniques—from device level up to the application level—need to work together to achieve the optimal lifetime and acceptable wearout levels at low cost. The notion of "cross-layer resilience" was first introduced to the computing community in the late 2000s [1], the key idea being to divide error and variation tolerance into a set of tasks, which can be implemented at different levels of the system stack as listed in Fig. 5.1. These resilience tasks can be treated as steps that the system follows to handle a particular reliability effect even as they may not occur sequentially [2, 3]. For example, cross-layer error predictions or detection for soft errors [3, 4], run-time sensing and actuation [5], etc. have been proposed to optimize the design of reliable systems. Orthogonal to these prior methods, the circadian rhythm-like accelerated and active recovery techniques demonstrated in this book are able to "repair" the wearout completely and efficiently by fully taking advantage of the unique BTI/EM recovery and frequency dependence behaviors; thus, they can be promising cross-layer solutions for further optimizing the system resilience. In this chapter, we present a set of solutions at different levels of the system hierarchy, with the circuit building blocks discussed in Chap. 4 serving as a key infrastructure for a cross-layer accelerated self-healing (CLASH) system which instruments the recovery from the circuit level to the system level. The CLASH system thus can operate for a longer time with higher performance in a refreshed mode, eventually leading to a significant reduction of required guardbands and better cumulative metrics (e.g., average performance). Since the proposed accelerated self-healing methods are orthogonal to previously proposed adaptive solutions, there are good

© Springer Nature Switzerland AG 2020
X. Guo, Mircea R. Stan, *Circadian Rhythms for Future Resilient Electronic Systems*, https://doi.org/10.1007/978-3-030-20051-0_5

Fig. 5.1 An illustration for
the concept of *cross-layer*
techniques

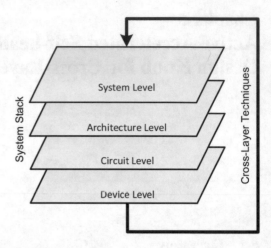

opportunities to combine them together to achieve a closer to optimal cross-layer resilience in a more effective way at lower cost. We also show that these CLASH techniques can provide an effective solution against both BTI and EM wearout for system designers, circuit designers, and the wider design community.

5.2   Accelerated and Active Recovery Design Space Exploration

In Chaps. 2 and 3, we have demonstrated experimentally that both BTI and EM wearout can be almost fully recovered by tuning knobs such as stress voltage/current, temperature, or both. When implemented on-chip, the tunability of these knobs, such as voltage range and temperature range, is somewhat limited; thus, we need first to perform a design space exploration of accelerated and active recovery under different on-chip scenarios, and second to find solutions for determining optimal operating schedules for both BTI and EM wearout. Details are presented in the following sections.

5.2.1   High Temperature or Negative Voltage? or Both?

As shown in Sects. 2.4 and 2.9, a high temperature of 110 °C and a negative voltage of −0.3 V when combined together demonstrate the effectiveness of the accelerated and active recovery techniques, as this leads to significant metric improvements, such as increased recovery rate, reduced design margin, and increased average performance. But it should be noted that the values of the temperature and negative voltage are not unique, in fact, due to the different application behaviors and

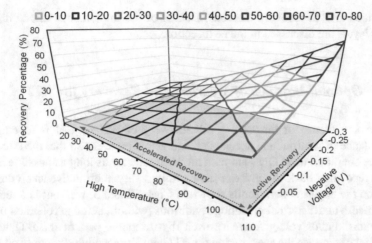

Fig. 5.2 BTI recovery rate under different combinations of high temperatures (accelerated recovery) and negative voltages (active recovery) when 6 h of recovery follow a 24 h accelerated stress

physical locations on chip, the available high temperatures and negative voltage resources might need to differ a lot. Thus it is necessary to explore other possible combinations with a range of temperatures and voltage settings in order to offer flexibility for cross-layer on chip implementations. Figure 5.2 plots predicted recovery rate under different combinations of high temperatures and negative voltages based on the analytic model detailed in Sect. 2.4.2. The same model parameters used in Sect. 2.7 are used here as well. We define "high temperature" as any temperature values above 20 °C and "negative voltage" as any voltage levels below 0 V. Assume that the chip has been stressed at nominal voltage under high temperature of 110 °C for 24 h; then, the recovery percentage of 72.3% demonstrated in Sect. 2.4 will become the upper limit for achievable recovery if the recovery voltages and temperatures are less than those max values. The surface plot presents the recovery percentage (recovered delay/net delay) increase due to wearout under various accelerated and active recovery conditions for a 6 h recovery period (long enough for reversible wearout to be fully recovered as shown in Fig. 2.23). It turns out that the same recovery rate can be achieved with multiple combinations of high temperatures and negative voltages. For example, to achieve a recovery percentage of ~30%, the combinations can be (50 °C, −0.3 V), (60 °C, −0.25 V), (70 °C, −0.2 V), (80 °C, −0.15 V), (90 °C, −0.1 V), (100 °C, −0.05 V), or (110 °C, 0 V). An essential observation is that an increase of 10 °C in temperature leads to similar recovery rate increase to that of a reduction of 0.05 V of the negative voltage value. These observations predict potential opportunities for "controlling" the recovery levels on chip via the available local voltages and heat. These results also indicate that different parts of the chip can achieve the same recovery rate even when the on-chip heat is not uniformly distributed or has fluctuations. Similar conclusions also apply to the EM accelerated and active self-healing techniques as they showed many similarities to BTI recovery methods. Details of the implementations at different system layers are discussed in the next sections.

5.2.2   Optimal Balance of Wearout and Recovery for BTI

Another key factor for ensuring a full recovery of BTI or EM is to employ a right balance of wearout and accelerated and active recovery so that the circuit can continue operating in an ON state with higher frequency as long as possible but can still be recovered back to the fresh state within a very short active sleep duration. Based on the experimental results shown in Chaps. 2 and 3, wearout is a relatively slow process under normal operating conditions (without being accelerated by high temperature and/or voltage/current); even the reversible part of the BTI wearout usually takes days, so the sleep period for BTI could be scheduled roughly on a daily (or several days) base—the accelerated and active recovery following a "diurnal" schedule of active sleep once every day (or even once every several days) being active.

The optimal scheduling strategy can be estimated based on the wearout and recovery models, or it can be explored during run time for better accuracy. To actually detect the optimal balance of active vs. sleep for a certain system, small embedded circuit-level wearout sensors discussed in Chap. 4 can be spread over the on-chip test structures, these sensors can feed the degradation information back to a controller which then enables the active accelerated recovery. Shown in Fig. 5.3 is a training-like process for BTI wearout: the accelerated and active recovery can be scheduled incrementally, e.g., every day, and then every other day, during the initial lifetime. The sensors monitor the degradation information correspondingly. When

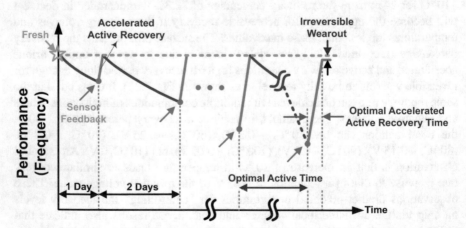

Fig. 5.3 Illustration of the "training" process for finding the optimal stress vs. recovery balance to fully recover from the BTI wearout effects

the system reaches the point where irreversible BTI wearout shows up, the optimal circadian rhythm will be defined as the last combination of stress and scheduled recovery duration. This process will be able to find exactly the optimal operating schedule for a reliability-critical system, and this schedule can also be integrated as part of a system-level scheduler for recovering from wearout during the rest of the system lifetime.

5.2.3 *Optimal Balance of Wearout and Recovery for EM*

Since EM-induced resistance increase starts after an extended time of stress accumulation, finding the optimal circadian rhythm for EM during run time has to be slightly different than for BTI. Illustrated in Fig. 5.4 is one suggested solution: in the early lifetime during the stress accumulation period, the EM sensors can be triggered rarely (e.g., once a month) to check the resistance until this starts increasing; then stress and recovery can be scheduled more similar to the BTI case, first on a daily basis, then every other day, and so on. The optimal stress and recovery schedule can be decided once the irreversible component of EM starts to be captured by the EM sensors. Compared to BTI, the overhead of finding the optimal schedule for EM can be smaller since even during active recovery periods the circuit blocks can continue operating if the PDN switching circuitry discussed in Sect. 4.3 is implemented on chip. The extended initial period of "constant resistance" also reduces the tracking overhead from the EM sensors.

In summary, this section and the last section provide several run-time solutions to find the exact operating schedule for full recovery. On one hand, such a schedule is useful for balancing the overhead of implementing recovery. On the other hand, the process of detecting this schedule increases the complexities of operating the system. Another, simpler scheduling solution we call "proactive recovery" is covered in Sect. 5.4.1.

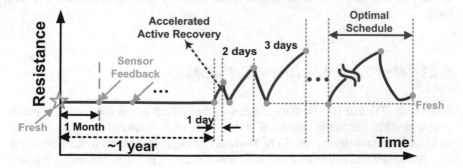

Fig. 5.4 Illustration of finding the optimal stress vs. recovery balance for EM during run time

5.3 Architecture-Level Accelerated Self-healing

In Chap. 4 we presented several circuit solutions for accelerated self-healing but such circuit-only solutions require fine-grained control and can introduce large overheads in some cases; dividing the recovery tasks across the system stack can be more efficient. In this section, we present several potential architecture-level accelerated healing techniques which can be employed to optimize the trade-offs and reduce the overhead.

5.3.1 Architectural Model for Wearout and Lifetime Analysis

Architecture research relies on abstract models that are used for capturing the lower level (circuit and device level) behaviors while eliminating unnecessary details; these models need to be "light" in terms of simulation complexity but still accurate. In this section, we use a recently developed open-source framework for architecture reliability called "OldSpot" [6] that is able to analyze wearout and estimate the lifetime of a system at the unit level with arbitrary distribution of workloads, and therefore variations of temperature, voltage, frequency, and so on, across time and space and arbitrary tolerance for failure. This model differs from previous tools by relaxing assumptions about the behavior and failure tolerance of the system and enabling fast lifetime and reliability modeling. OldSpot is compatible with other architecture-level frameworks, e.g., it can receive performance data from gem5 [7], power data from McPAT [8], temperature data from HotSpot [9], and a floorplan created with ArchFP [10]. The tool then computes per-unit wearout rates and runs Monte Carlo simulations to output reliability distributions for each unit and the overall system. This tool is able to help an architect predict the impact of expected workloads on a design in the early design phase and estimate which regions of a chip are experiencing wearout "hot spots" and should be targeted for aging mitigation techniques such as the accelerated and active recovery techniques discussed in this book.

5.3.2 Unit-Level Accelerated Self-healing

Since both BTI and EM wearout depend on temperature, voltages, and switching activities, this inevitably means that different units experience very different wearout behaviors during run time. There are several alternatives of instrumenting the accelerated and active recovery solutions discussed in Chap. 4: the first one is a coarse-grained solution where wearout sensors and recovery circuitry are distributed evenly at the core level so that the whole core recovers in a similar way—the benefit of this solution is that it is relatively easy to control and instrument, but the downside is that it means that some units which experienced more wearout may not get full recovery, while other, less stressed units, may become over-recovered, thus the overall benefits of recovery becoming suboptimal. A more effective (and more economical) solution is a fine-grained *unit-level accelerated self-healing*, in which "wearout hotspots" or wearout-critical units are predicted by architectural tools, and the accelerated self-healing is only instrumented and applied to these units. With a pre-RTL reliability simulator such as OldSpot, we are able to conduct an analysis of such methods. Figure 5.5 shows an example where we run a benchmark (*Cholesky*) from the PARSEC suite [11] on Intel's Nehalem architecture with parameters shown in Table 5.1. Figure 5.5a shows the heat map from HotSpot and Fig. 5.5b shows the corresponding wearout map from OldSpot. Here wearout is mainly dominated by BTI and EM effects. The results show that both wearout mechanisms are not uniform across the whole chip, with some functional units experiencing more wearout due to higher switching activities that lead to higher power consumption and higher temperatures; thus, recovery solutions should be instrumented differently across the chip. For instance, based on the results from Fig. 5.5b, reorder buffers (rob) experience more wearout; thus, recovery circuitry such as power gating and on-chip heaters should be firstly instrumented on these rob units; as for units such as the L2 cache for which wearout has a relatively smaller impact, it is less imperative to add additional units to accelerate their recovery as purely passive recovery could

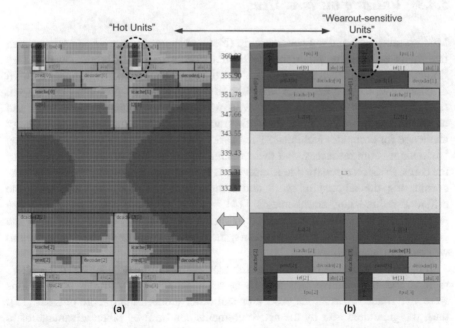

(a)   (b)

Fig. 5.5 (**a**) Heat map representing the average temperature of each unit when running *Cholesky* [11]; (**b**) a corresponding wearout map representing relative wearout rates of each unit. In (**a**), red indicates hotter temperatures and blue indicates cooler ones. In (**b**), red indicates faster wearout, while blue indicates slower wearout

Table 5.1 Simulated system parameters for OldSpot

Parameter	Value
Technology node	65 nm
Instruction set architecture (ISA)	×86
Microarchitecture	Nehalem
Supply voltage	1.1 V
# of cores	4
CPU clock frequency	2.66 GHz
Instruction cache (ICache)	32 kB
Data cache (DCache)	32 kB
Private L2 cache	256 kB
Shared L3 cache	8 MB

be enough for such units. In summary, unit-level accelerated self-healing provides fine-grained control at the functional block level, and is potentially more effective because the functional units that suffer from BTI and EM wearout the most can be instrumented with more recovery resources. This fine-grained method can also be less costly in terms of hardware overhead compared to the coarse-grained solution.

5.3.3   Utilizing Intrinsic Heat

Modern multicore and network-on-chip (NoC) systems can have hundreds or even thousands of cores or functional units, but due to thermal design power (TDP) limitations, not all the cores can operate at the same time, and a significant amount of cores need to be OFF—this was dubbed the *dark silicon* problem [12]. Recent research [13] has pointed out that even with the latest FinFET technology and novel processor architectures, the dark silicon issue still remains a major design challenge for computer architects. The "dark" part of the chip usually leads to some "redundant" core resources, and these resources can be a single core or a subset of the cores. Prior research tried to leverage this redundancy by developing frameworks considering the amount of work and temperature variations and analyzing the different redundancy arrangements [14]; on top of these techniques, if these resources are scheduled and allocated in such a way that they can be healed by the generated heat from the active elements, the average lifetime of the whole system will be significantly improved.

It has been shown in Fig. 5.5a that the temperature difference between the active regions and the inactive regions can be as large as 30 °C; similar observations were also made on an Alpha processor and other modern many-core designs [15]; thus, the generated heat by the active elements can be fully taken advantage of to heat up the idle units that need to be rejuvenated. Figure 5.6 illustrates a potential implementation in a simplified multicore system, where the sleeping cores (or units) are always surrounded by active cores (or units); by switching the workload among the cores, these sleeping cores can be deeply rejuvenated by the heat generated

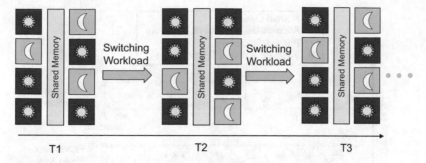

Fig. 5.6 A potential self-healing solution in a multicore system. Dark silicon and core redundancy can be utilized to improve the lifetime of the whole system. $T1$, $T2$, and $T3$ are time points during the lifetime

by the active cores. This method is application-dependent and also needs a system scheduler to be able to make the recovery control decisions. There will also be some workload migration overhead, but this solution that takes advantage of the inherent dark silicon behavior doesn't need to stall the system completely during recovery, and it also eliminates the power overhead introduced by circuit-level on-chip heaters (presented in Chap. 4). Scheduling details are discussed in Sect. 5.4, and a potential cross-layer implementation is covered in Sect. 5.6.

5.4 Scheduling at the System Level

The right balance of wearout and recovery can lead to almost full recovery for both BTI and EM recovery—this has inspired the idea of active recovery at the system level, for which a task scheduler can implement scheduling based on desired "circadian rhythms"; this section briefly discusses several ways of scheduling and how to apply these for different applications.

5.4.1 Reactive Recovery vs. Proactive Recovery vs. AC Stress

Since we know that recovery can be used to deeply rejuvenate the system from wearout, it may be worthwhile sometimes to schedule the recovery periods *proactively*. During these proactive recovery periods several accelerated and active recovery techniques can be applied, but it is important to decide when to insert these recovery periods in order to avoid the irreversible component of wearout. Philosophically, there are two alternatives for scheduling recovery: reactive, when a system or part of system has aged by a particular threshold amount, and proactive, in anticipation of future wearout. Figure 5.7 illustrates the differences. In the reactive

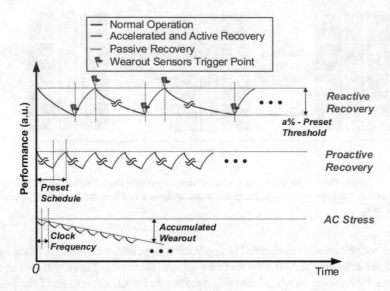

Fig. 5.7 Illustration of reactive recovery vs. proactive recovery vs. AC stress

recovery case, a wearout threshold $a\%$ is preset as the upper limit and starting point for the recovery periods. This threshold can be pre-calculated based on the wearout models or simply can be the design margin. During operation, wearout sensors need to be able to track and feed the wearout information back to the scheduler frequently, as the time to reach the threshold is unpredictable due to the different switching behaviors during different periods of lifetime; thus in this case, wearout sensors are very important and critical for deciding when to start recovery. As for the proactive recovery case, the recovery schedule can be preset, and this schedule needs to be set such that the irreversible components don't start to accumulate; this can be determined based on pre-RTL space exploration (e.g., with OldSpot) and wearout models which are usually provided by the foundry. Once this preset schedule is determined, the scheduler inserts the recovery periods proactively so that recovery can always compensate the accumulated wearout effects. In this case, wearout sensors are still necessary, but just as a sanity check. Proactive recovery acts similar to a "slow AC stress" with a *skewed* duty cycle, in which recovery and stress appear as pairs. Note that it is related but not exactly identical to normal AC stress (defined in Sect. 2.6.1) which is also shown in the figure. AC stress is when there are switching activities and the transistors/metal wires are under periodic stress and *passive* recovery only; wearout still accumulates in this case as the passive recovery is too slow for fully compensating for the stress.

Reactive accelerated recovery seems more "economic" since it is only scheduled when needed, but it needs to track changing wearout levels (such as threshold voltage increase or resistance increase), is unpredictable, thus potentially introducing performance and/or energy overheads at inopportune times, and likely leads to a smaller improvement in overall lifetime as it accumulates more irreversible aging upfront, thus leading to a lower expected performance and energy—the circuit operates more time in a stressed mode. In addition, the preset wearout threshold is hard to determine. Proactive recovery on the other hand is easier to implement, the recovery periods are scheduled ahead of any sign of wearout resulting in the system operating for longer time in a "refreshed" mode, thus leading to better overall performance and lifetime, and has better cumulative metrics as well. Most importantly, the preset schedule can even follow users' schedules and be fully compatible with human "circadian rhythms"; details are discussed in the next section.

5.4.2 Application-Dependent Proactive Recovery with Scheduling

Based on the experimental results and analysis in Sects. 2.9.2, 3.6, and 5.2, wearout is a relatively slow process under normal operating conditions. For example, even the reversible part of the BTI wearout usually takes at least 1 day under the constant stress (e.g., 31 h), so the sleep period could be predefined on a daily (or several days) base—the accelerated and active recovery being able to operate on a 1 day (or several days) period of being active. In this section, we discuss two use cases and the potential circadian rhythms for each of them.

- *Mobile or Edge Devices*—For applications such as mobile devices, wearables, or consumer electronics, circadian rhythms for recovery can ideally follow human users' active vs. sleep patterns. For example, these devices are mostly used during daytime, and this active time can be scheduled as (say) 12–16 h. During the night, most of these devices can be in OFF state that corresponds to human sleep duration. Although certain functional units on the chip need to be active for state retention, many wearout-critical units can be recovered with the fine-grained accelerated self-healing techniques we discussed in the previous sections. The duration of such recovery depends upon the sleep conditions, i.e., available temperatures and voltages. Assume that the system is active for 12–16 h under normal operating conditions (room temperature and nominal voltage). The worst-case scenario is when transistors are always under constant stress (DC stress), but the resulted BTI wearout is still within the reversible region (e.g., within 31 h). Based on the model predictions in Sect. 2.4.2 and the analysis in Sect. 2.9.2, the required time for a full BTI recovery under different sleep conditions is plotted in Fig. 5.8. It shows that recovery is very fast and needs about 23.2 min under the combinations of very high temperature of 110 °C and −0.3 V, which refers to the condition we conducted the recovery measurements. But on a real operating chip, it is clear that these temperatures or negative voltages might not be always available unless we instrument the heaters or negative voltage generator. However, this plot shows that even when a slightly higher temperature or negative voltage is applied, the recovery can still be much faster than for

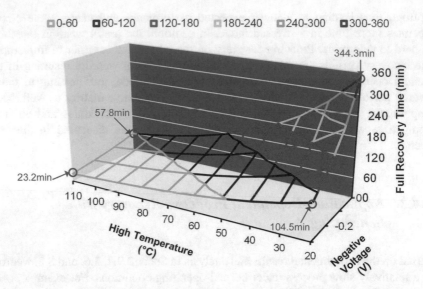

Fig. 5.8 Recovery time under different accelerated and active recovery conditions after 12 h constant stress under regular operation condition (room temperature, nominal V_{dd})

regular operating condition. For example, for 50 °C and −0.3 V, the system needs only about 57 min to achieve full BTI recovery. Such devices could benefit from such short sleep intervals when part of them are put in sleep states.

- *High Performance Computing (HPC)*—For data center or server applications that are ON most of their lifetimes, it will not be feasible to fully turn off the system as frequently as for mobile devices or embedded systems mainly because of the high setup times and full usage of the system. However, recent research [16–18] have proposed energy-efficient solutions that implement novel load balancing and/or scheduling algorithms so that the idle and lightly loaded cores are able to be switched to sleep states. For example, in [17], the authors introduced a dynamic power management policy called *SoftReactive*, which is able to put servers to sleep in a conservative way by setting a wait time when the workload drops; then, when the load increases back, the policy quickly turns servers back. This work demonstrates that as the scale of data center increases, the effectiveness of sleep states is even more pronounced because the setup time has less and less effect on performance. During these intrinsic sleep state, circuit level and architecture level the active accelerated self-healing techniques discussed in the previous sections can work together to ensure full recovery capabilities. Larger data centers can exploit more benefits of sleep states for recovery even by employing a simple dynamic power management policy.

5.5   Recovery-Driven Design Methodology for Resilient System Design

It has been shown in the previous sections that active accelerated self-healing can be achieved at different levels of a system stack during run time, but the hardware blocks and all the scheduling decisions need to be instrumented during the design phase—for example, where to place the sensors, how many of the recovery circuitry are required, what will be the trade-offs, etc. In this section, we discuss a potential design methodology that is driven by recovery.

Figure 5.9 illustrates the recovery-driven design flow. It starts from the design specifications (SPEC) and applications which define the expected lifetime, clock frequency, power and area budget, and so on. This flow assumes that high-level architecture-level decisions based on the SPEC have been made; these decisions can be type of ISA, number of cores, and so on. This information can be used as the input for pre-RTL and architecture-level simulation tools to estimate the power, thermal, and performance behaviors. Reliability tools such as OldSpot discussed in Sect. 5.3.1 that are based on wearout models can take the reported behaviors from such tools and determine the wearout behavior at a higher level, such as which units could be potentially wearout-critical and what would be the possible operating schedules for the proactive recovery. These decisions can guide the RTL design

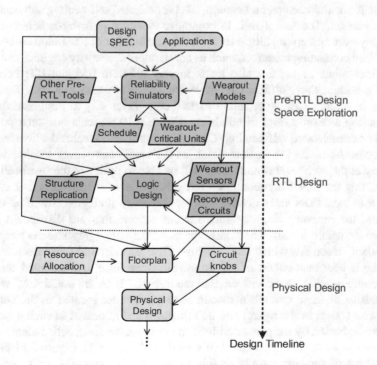

Fig. 5.9 An illustration of recovery-driven design methodology in an IC design cycle

phase when designing the microarchitecture; for example, more wearout sensors can be instrumented for the wearout-critical units, multiple copies of these units can be integrated in the design, and so on. The control logic for closing the wearout and recovery loop is also implemented in this stage of the design process. During the physical design stage, core/resource allocation solutions that are beneficial to accelerated recovery (discussed in Sect. 5.3.3) and PDN design against wearout (discussed in Sect. 4.3) can be part of the floorplan decisions. Similarly, recovery circuitry such as on-chip heating elements or sensors can be placed close to the wearout-critical units. During the rest of the physical design steps, circuit knobs that affect the recovery levels also need to be considered; examples of such knobs are the logic depth (discussed in Sect. 4.2.2) or power gating styles (discussed in Sect. 4.2.3). As reliability is usually not addressed in every design stage in most of the current design flows, this section gives an example of considering reliability during the whole design process, and it contributes to the efforts of developing future design methodologies for ensuring reliability.

5.6 Putting It All Together—CLASH: Cross-Layer Accelerated Self-healing System

So far, circuit, architecture, and system-level accelerated self-healing solutions have been discussed. The recent shift in computer architecture towards heterogeneous and many-core systems significantly increases the number of integrated cores, while specialized computing resources, such as accelerators or security engines, can speed up various applications; this also leads to very different EM and BTI behaviors across the chip; thus, different recovery strategies are also required. Designers need to integrate single-layer solutions in a cross-layer way to minimize the cost for ensuring a desired level of reliability. Figure 5.10 presents one such potential cross-layer accelerated self-healing (CLASH) system with localized active recovery techniques. As discussed in Sect. 5.3.2, at the architecture-level localized active recovery at the core level or unit level will be able to optimize the overhead while rejuvenating the "aged" system. Each red/yellow square represents a core or logic block with local PDN and can have different recovery strategies. The blue squares represent the recovery circuitry and wearout sensors that are distributed across the core or within a unit. The "dark" parts of the chip (gray squares) represent "redundant" resources which have intrinsic OFF periods, and these resources can be a single core or a subset of the cores. Since we have demonstrated that high temperature is able to accelerate the recovery of both BTI and EM wearout mechanisms, if these redundant resources (e.g., the core located in the center of the system shown in the figure) can be scheduled and allocated in such a way that they can be healed by the generated heat from the neighboring active elements, the recovery can be further accelerated (discussed in Sect. 5.3.3). Figure 5.11 presents an example of run-time proactive scheduling for BTI and EM active recovery.

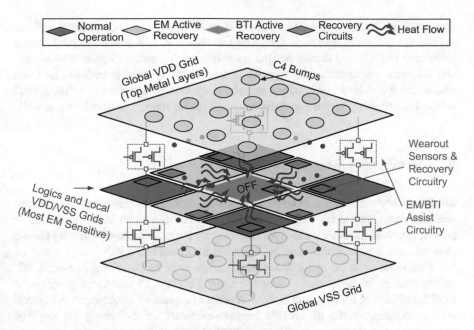

Fig. 5.10 An illustration of cross-layer accelerated self-healing (CLASH) system

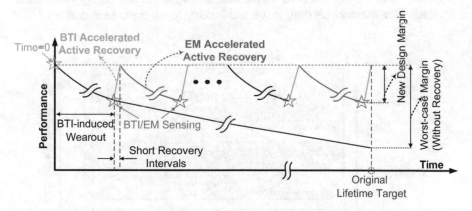

Fig. 5.11 Illustration of periodic proactively scheduled EM/BTI recovery

In the early lifetime, since EM-induced stress hasn't reached the nucleation threshold, theperformance degradation will be caused mainly by BTI; novel BTI and EM sensors can be deployed to track wearout and check the degradation information. Short intervals of BTI active recovery periods are inserted proactively to bring the chip back to the fresh state in time; during these intervals, certain functional units need to be in retention mode, alternatively, their workload can be shifted to other redundant resources. EM active recovery periods can be scheduled either from when the void nucleation happens or even earlier. Based on the measurement results presented in Chap. 3, early EM recovery is more economic and efficient, and the system can still be in operation during EM recovery, so EM active recovery can be scheduled periodically during normal operation with a small switching overhead. Overall, such a scheduling strategy can potentially fully recover both the BTI and EM wearout and lead to a situation where the system always runs in a "refreshed" mode; the necessary wearout guardbands can then be significantly reduced as well.

5.7 Trade-off Discussions

Similar to any other added features in hardware design, instrumenting accelerated and active recovery techniques with hardware is not free either in terms of area or power. However, through cross-layer implementations, the trade-offs among power, performance, and reliability can be balanced in an almost optimal way. Some of these trade-offs can be optimized during design time as described in Sect. 5.4.2, others need to be taken care during run time. Figure 5.12 illustrates an example implementation optimizing the trade-offs across layers. For example, at the circuit level, generating heat with on-chip heating elements, or delivering the negative voltages by the circuit-level voltage generator, both will cause additional power overhead during sleep intervals which can be expensive for energy-constrained systems. To minimize this cost, at the architecture level, the core allocator or load

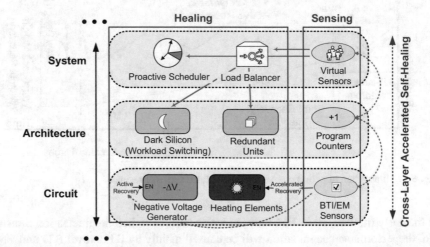

Fig. 5.12 Cross-layer implementation of accelerated self-healing for leveraging the trade-offs. At the circuit level, recovery circuit and wearout sensors are distributed on the wearout-critical units, and they are triggered by higher level decisions from scheduler or load balancer. Architecture-level accelerated self-healing solutions can utilize some intrinsic sleep behaviors and heat to recover inactive parts. It can also compensate some of the power overhead introduced by the circuit-level recovery solutions. System-level scheduling is able to divide the recovery tasks and make the high-level recovery decisions

balancer can first work together to assist the process by shifting applications among different cores and allocating the resources in an optimal way (e.g., as shown in Fig. 5.6) so that the circuit-level solutions can stay inactive, thus avoiding the extra power consumption. At higher levels of the stack, the system scheduler can optimize the trade-offs by scheduling the necessary accelerated self-healing periods proactively (e.g., following a pattern as discussed in Sect. 5.4.1). Sensing elements are the key elements that enable transparency between layers; such sensors can be physical sensors (BTI sensors detailed in Sect. 4.4 or EM sensors detailed in Sect. 4.5) at the circuit level, program counters at the architecture level, or virtual sensors at the system or application levels. This suggests that hardware and software need to work together to take full advantage of the recovery behaviors, and the collaborative efforts from all layers can potentially ensure that the whole system will always be rejuvenated from stress so that it can run reliably over the entire lifetime with a relatively low cost.

5.8 Conclusions

Since implementing accelerated self-healing at a single layer can potentially lead to large power and area overheads we discussed several cross-layer implementations of accelerated and active recovery strategies for both BTI and EM. As highlighted in Fig. 5.13, we discussed both architecture and system-level recovery solutions that can benefit from the active accelerated self-healing mechanisms presented in the previous chapters. Distributing the recovery tasks across the system stack can improve performance, and reduce power and area costs by taking advantage of the characteristics of each layer. We also suggested at a high level how recovery can be integrated as part of the design decisions in a conventional design cycle. Overall, active accelerated self-healing is a promising technique for solving wearout issues and can be introduced as a key design knob for cross-layer resilience during the design process to achieve the optimal resilience effectively.

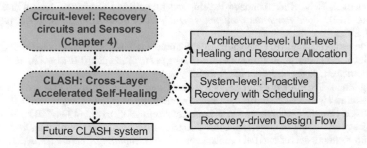

Fig. 5.13 Chapter 5 highlights

References

1. Computing Community Consortium (CCC) Visioning Study on Cross-Layer Reliability. http://www.relxlayer.org/.
2. Nicholas P Carter, Helia Naeimi, and Donald S Gardner. Design techniques for cross-layer resilience. In *Proceedings of the Conference on Design, Automation and Test in Europe*, pages 1023–1028. European Design and Automation Association, 2010.
3. Subhasish Mitra, Kevin Brelsford, and Pia N Sanda. Cross-layer resilience challenges: Metrics and optimization. In *Design, Automation & Test in Europe Conference & Exhibition (DATE), 2010*, pages 1029–1034. IEEE, 2010.
4. E. Cheng, J. Abraham, P. Bose, A. Buyuktosunoglu, K. Campbell, D. Chen, C. Y. Cher, H. Cho, B. Le, K. Lilja, S. Mirkhani, K. Skadron, M. Stan, L. Szafaryn, C. Vezyrtzis, and S. Mitra. Cross-layer resilience in low-voltage digital systems: Key insights. In *2017 IEEE International Conference on Computer Design (ICCD)*, pages 593–596, Nov 2017.
5. S Sarma, N Dutt, N Venkatasubramanian, A Nicolau, and P Gupta. Cyberphysical system-on-chip (cpsoc): Sensor actuator rich self-aware computational platform. *University of California Irvine, Tech. Rep. CECS TR-13-06*, 2013.
6. Alec Roelke, Xinfei Guo, and Mircea R Stan. OldSpot: A Pre-RTL Model for Fine-grained Aging and Lifetime Optimization. In *Computer Design (ICCD), 2018 IEEE International Conference on*. IEEE, 2018.
7. Nathan Binkert, Bradford Beckmann, Gabriel Black, Steven K Reinhardt, Ali Saidi, Arkaprava Basu, Joel Hestness, Derek R Hower, Tushar Krishna, Somayeh Sardashti, et al. The gem5 simulator. *ACM SIGARCH Computer Architecture News*, 39(2):1–7, 2011.
8. Sheng Li, Jung Ho Ahn, Richard D Strong, Jay B Brockman, Dean M Tullsen, and Norman P Jouppi. McPAT: an integrated power, area, and timing modeling framework for multicore and many core architectures. In *Microarchitecture, 2009. MICRO-42. 42nd Annual IEEE/ACM International Symposium on*, pages 469–480. IEEE, 2009.
9. Wei Huang, Shougata Ghosh, Sivakumar Velusamy, Karthik Sankaranarayanan, Kevin Skadron, and Mircea R Stan. HotSpot: A compact thermal modeling methodology for early-stage VLSI design. *IEEE Transactions on Very Large Scale Integration (VLSI) Systems*, 14(5):501–513, 2006.
10. Gregory G Faust, Runjie Zhang, Kevin Skadron, Mircea R Stan, and Brett H Meyer. ArchFP: Rapid prototyping of pre-RTL floorplans. In *VLSI and System-on-Chip (VLSI-SoC), 2012 IEEE/IFIP 20th International Conference on*, pages 183–188. IEEE, 2012.
11. Christian Bienia. *Benchmarking modern multiprocessors*. Princeton University, 2011.
12. Hadi Esmaeilzadeh, Emily Blem, Renee St Amant, Karthikeyan Sankaralingam, and Doug Burger. Dark silicon and the end of multicore scaling. In *Computer Architecture (ISCA), 2011 38th Annual International Symposium on*, pages 365–376. IEEE, 2011.
13. Jorg Henkel, Heba Khdr, Santiago Pagani, and Muhammad Shafique. New trends in dark silicon. In *Design Automation Conference (DAC), 2015 52nd ACM/EDAC/IEEE*, pages 1–6. IEEE, 2015.
14. Lin Huang and Qiang Xu. Characterizing the lifetime reliability of many core processors with core-level redundancy. In *Computer-Aided Design (ICCAD), 2010 IEEE/ACM International Conference on*, pages 680–685. IEEE, 2010.
15. Cheng Zhuo, Kaviraj Chopra, Dennis Sylvester, and David Blaauw. Process variation and temperature-aware full chip oxide breakdown reliability analysis. *Computer-Aided Design of Integrated Circuits and Systems, IEEE Transactions on*, 30(9):1321–1334, 2011.
16. Paul Bogdan, Siddharth Garg, and Umit Y Ogras. Energy-efficient computing from systems-on-chip to micro-server and data centers. In *Green Computing Conference and Sustainable Computing Conference (IGSC), 2015 Sixth International*, pages 1–6. IEEE, 2015.

17. Anshul Gandhi, Mor Harchol-Balter, and Michael A Kozuch. Are sleep states effective in data centers? In *Green Computing Conference (IGCC), 2012 International*, pages 1–10. IEEE, 2012.
18. A. Paya and D. Marinescu. Energy-aware load balancing and application scaling for the cloud ecosystem. *IEEE Transactions on Cloud Computing*, PP(99):1–1, 2015.

Part IV
Wearout and Recovery in Advanced Technology and Emerging Applications

Chapter 6
Design and Aging Challenges in FinFET Circuits and Internet of Things (IoT) Applications

6.1 Overview

Over the past few decades, there have been continuous speculations about the demise of Moore's law [1] due to many reasons, including lithography, the limitations of MOS planar devices, and increased manufacturing complexity. FinFET technologies have been introduced to enable further increases in the levels of integration, and they enabled scaling below 20 nm thus helping to extend Moore's law by a precious decade with another decade very likely in the future when scaling to 5 nm and below. Due to superior electrical parameters and unique structure, these 3D devices offer significant performance improvements and power reductions compared to planar CMOS. As the semiconductor industry is entering into the sub-10 nm era, FinFETs have become dominant in most of the high-end products; they are also an ideal candidate for low power and energy-constrained applications, of which Internet of Things (IoT) has grown significantly in recent years. IoT introduces a paradigm where humans and "things" are interconnected; IoT is a powerful enabler that makes technology more human centric and real time. Although the IoT industry has not fully migrated to deeply scaled technologies because of cost and leakage issues, recent advances in technology such as FinFET have demonstrated a compelling combination of performance, power, highest integration, and ease of design for low power. This has pushed a new wave for IoT industry to switch to these advanced technology nodes [2].

In both industry and academia, the transition of technology from planar CMOS devices to FinFET devices is still ongoing. Within this context, it is critical for digital circuit designers and researchers to fully understand the fundamental challenges

© Springer Nature Switzerland AG 2020
X. Guo, Mircea R. Stan, *Circadian Rhythms for Future Resilient Electronic Systems*, https://doi.org/10.1007/978-3-030-20051-0_6

and opportunities introduced by the advanced technology nodes. In the first part of this chapter we present a comprehensive study of these aspects from device to circuit levels; this is followed by comparisons across multiple technology nodes ranging from conventional bulk to advanced planar technology nodes such as fully depleted silicon-on-insulator (FDSOI), to FinFETs. To conduct simulations, we use both state-of-the-art industry-standard technology models for current nodes and also predictive models for future nodes. The study shows that besides the well-known performance and power benefits, FinFET devices also exhibit significant reduction of short-channel effects and extremely low leakage, and many of the electrical characteristics are also close to ideal as in old long-channel abstractions; FinFETs seem to have acted as a reset of technology scaling trends. However, the combination of new device 3D structures, double/multi-patterning, many more complex rules, and unique thermal/reliability behaviors have created more new technical challenges. Adapting to the new challenges and fully benefiting from FinFETs in design flows will require growing knowledge and design experiences, and this part of our work aims to add to that knowledge base.

From an application perspective, the reliability of the IoT devices becomes extremely critical. Circuit aging[1] discussed in this book will have a direct impact on the lifetime of these devices and their performance. It is necessary to understand how, and on what level, aging affects the IoT chips for different categories of current and future IoT applications. In the previous chapters, we mainly investigated the underlying mechanisms of circuit aging and recovery behaviors, in the second part of this chapter, we look into the impact of circuit aging within the context of applications, especially in the IoT domain. We answer the above questions by conducting extensive circuit-level simulations with foundry-calibrated transistor aging models in an advanced FinFET node. Since aging is highly dependent on application behavior that defines the operating voltage, the temperature, and active time, we first perform a survey of existing IoT applications and classify them based on aging-related metrics. By studying aging behaviors in each category, we confirm that aging can impose large degradations in circuit performance and design margins for several types of IoT applications. Our results strongly suggest that designers need to take both battery lifetime (constrained by energy-efficiency) and chip lifetime (constrained by circuit aging) into consideration in a comprehensive design flow. As flat guardband approaches can introduce unacceptable energy overheads for IoT, several dynamic solutions that are able to benefit from the recovery behaviors and the corresponding circuit solutions introduced in the previous chapters are also presented.

---

[1]This chapter focuses mostly on transistor aging (especially BTI and HCI). Battery aging, socket (and holder solder) aging in IoT devices are also important, but they are beyond the scope of our discussion.

6.2 Back to the Future: Digital Circuit Design in the FinFET Era

6.2.1 Motivations for Studying FinFET Technology

Although continuous scaling of planar CMOS devices has delivered increasing performance and transistor densities, it reached a point where increased leakage current, fluctuation of device characteristics, and short-channel effects became serious obstacles for further scaling. This was mainly due to the fact that in deeply scaled planar devices the gate loses the ability to fully control the channel; this led to transistors that were never fully turned off and leaked continuously. To address this, gate oxides were aggressively thinned and high-k dielectric gate materials were adopted to increase the gate-channel capacitance. However, the gate-related issues, such as gate leakage and gate-induced drain leakage (GIDL) increased [3, 4]. FinFET devices became attractive candidates for sub-30 nm nodes because of their unique channel structures with much better gate control that enables a much improved short-channel control, thus only little or even no doping is needed in the channel [5, 6]. The transistor threshold voltage V_{th} can be further scaled down in FinFETs for both improved device performance and much lower operation voltages. Lower channel doping also reduces dopant ion scattering, thus leading to better drive currents and mitigates random dopant fluctuations (RDF) effects [7–9]. FinFETs back-end-of-line (BEOL) fabrication is fully compatible with planar devices in both bulk and SOI varieties, and this reduces the need for any new and FinFET-specific developments for interconnect. However, the advent of FinFETs has also brought a few changes and challenges to digital circuit design due to their unique gate structure and electrical properties. This has also affected circuit design decisions and introduced several new design trade-offs. For example, FinFET devices have a significant amount of parasitics that need to be modeled precisely and be carefully addressed in physical design; this is especially pronounced in SRAM and analog circuits. In addition, new circuit techniques are needed for threshold voltage modulation and memory read/write assist in SRAMs to replace techniques (e.g., body biasing) that worked well in planar but are inefficient or unavailable for FinFET. The necessary double- and multi-patterning also requires more effort from tool vendors and designers to work together to make sure the layout coloring is correct (colors refer to the different exposures of the same layer while performing double/multi-patterning). New design constraints have also been added in FinFET circuit design, e.g., width quantization and self-heating effects, which designers need to address in early design phases. In this chapter, we study these aspects from both a device and a circuit design perspective. To study these challenges, we simulate across multiple technology nodes which cover a wide range of gate lengths and also substrates, including FD-SOI and bulk. For FinFETs, we simulate with a

$1\times$ nm industry-standard node[2] and a 7 nm predictive node [10]. We restrict our focus to digital circuits in this book, but several of the findings can be applied for analog design as well.

6.2.2 FinFET Scaling and Sizing

Compared to conventional planar devices (either bulk or SOI substrates), FinFET devices employ unique 3D gate structures that enable some special properties for FinFET circuit design. Figure 6.1 shows a planar device and a FinFET device (the substrate is not included in the figure). While the channel of a planar device is horizontal, the FinFET channel is a thin vertical "fin" with the gate fully "wrapped" around the channel formed between the source and the drain. The current flows parallel to the die plane whereas the conducting channel is formed around the edges of the fin. With this structure, the gate is able to fully deplete the channel so that it has much better electrostatic control over the channel.

FinFETs have several varieties and are usually classified by gate structure or type of substrate. Based on differences among gate structures, there are two types of FinFETs—shorted-gate (SG) FinFETs and independent-gate (IG) FinFETs. In SG devices, the left and right sides are connected together in a wraparound structure as shown in Fig. 6.1; this type can serve as a direct replacement for the planar devices which also have one gate, a source, and a drain (three-terminal devices). In IG FinFETs, the top part of the wraparound gate structure is etched out and this results in two separate left and right sides that can act as independent gates and can

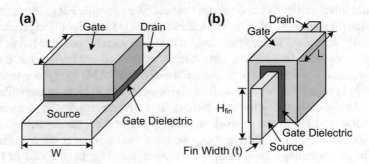

Fig. 6.1 Illustration of structural differences (no substrate): (**a**) planar device; (**b**) FinFET device

---

[2]In advanced technology nodes the "numbering" scheme is somewhat "fuzzy"; while in older technologies the node "number" used to denote the smallest feature size—usually the transistor gate length, in modern technologies the node number does not refer to any one feature in the process, and foundries use slightly different conventions. In this chapter, we use $1\times$ to denote the 14–16 nm FinFET nodes offered by several foundries.

Fig. 6.2 Cross-sectional
view of structural differences
between (**a**) bulk FinFET and
(**b**) SOI FinFET

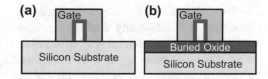

be controlled separately (four-terminal devices) [11, 12]. Although IG FinFETs offer
more design options, the fabrication costs are also higher in general, and commercial
foundries are not usually offering these as an option. Depending on the type of
substrate, the FinFETs can be either SOI or bulk FinFETs as shown in Fig. 6.2.
The fabrication of both types of FinFET devices is compatible with those of the
conventional planar devices fabricated on either bulk or SOI wafers. Similar to the
argument in planar CMOS devices, SOI FinFETs that are built on SOI wafers have
a lower parasitic capacitance and slightly less leakage. However, bulk FinFETs are
more familiar to designers, fabrication costs are lower, and they also have better
heat transfer to the substrate compared to SOI FinFETs [11], thus bulk FinFETs are
usually preferred for most of the digital applications. For these reasons, in this book
we also mainly focus on bulk FinFETs.

Unlike planar technologies for which the transistor width is a continuous value
fully under the control of designers, for FinFETs device widths are quantized
into units of whole fins. The effective gate width of a FinFET device is roughly
$n \cdot (2H_{fin} + t)$, where n is the number of fins, t is the fin width, and H_{fin} is
the fin height as illustrated in Fig. 6.1b. Since the gate of a FinFET device is
designed to achieve good electrostatic control over the channel, and because of
the etching uniformity requirements, the fin dimensions (e.g., height H_{fin}) are not
under designers' control, and thus the device width can't have an arbitrary value
as in planar transistors. Wider transistors with higher on-currents are obtained by
using multiple fins, but the range of choices is limited to integer values only. This is
known as the *width quantization* issue [13–15]. The quantization issue doesn't allow
flexibility in terms of device sizing which becomes problematic especially in SRAM
and analog designs. Designers need to adapt to this new constraint when designing
circuits [16]. An alternative solution is for the foundry to provide the designers
with multiple choices of fin heights [17]. For example, [18] did an early attempt by
exploring the design space of FinFETs with double fin heights and showed that the
lack of continuous sizing can be somewhat compensated; this method though has
many uncertainties from both fabrication costs and manufacturing difficulties, so
it is unlikely to become widely available. In summary, for digital circuits, width
quantization is inconvenient but is not be a big impediment since most of the
cells can be designed and sized with the limited choices of device widths that are
available.

As fin height H_{fin} determines the overall width of a device, this becomes a very
important parameter for circuit designers but they don't actually have control over it.
Smaller fin heights offer more flexibility in terms of sizing, but also lead to needing
more fins for wider transistors, which means more silicon area. In contrast, FinFET

devices with taller fins offer less flexibility in sizing but have a smaller silicon footprint. It is interesting that *the increasing fin heights for successive FinFET nodes along with the lateral scaling are able to actually accelerate "Moore's law"-style scaling in some sense, Intel calls this "hyperscaling,"* but higher fins might also result in larger short-channel effects and more structural instabilities [3, 11]. In addition, taller devices will also lead to an increase in unwanted coupling and fringing capacitances. These trade-offs indicate some potential opportunities for device-circuit codesign that are unlikely to become available to fabless companies but can be important for vertically integrated companies which have access to their own fabs. An example of such codesign is to explore the design space of current (drive strength) versus capacitance for various possible fin heights. As the technology node already approaches the sub-10 nm range, such a study is becoming increasingly important as fabrication difficulties keep increasing, which can also result in drastic changes in design trade-offs [19].

6.2.3 FinFET Fabrication

The fabrication of FinFETs is relatively straightforward and compatible with the conventional planar device fabrication process [20], but there are still new challenges, such as fin shape control and recess of shallow trench isolation (STI) oxide which are critical in the integration of FinFETs. This section lists several of such fabrication advances along with the associated challenges in the FinFET era.

6.2.3.1 Double/Multi-Patterning

Although CMOS transistor keeps scaling to the order of a few nanometers, lithography still uses 193 nm wavelength light, which makes printability and manufacturability extremely challenging, so much so that beyond 20 nm the use of multi-patterning is required for device fabrication [21]. By using multi-patterning, a single layer is decomposed into two or more masks and manufactured through two or more exposure steps as shown in Fig. 6.3. These masks are then combined to

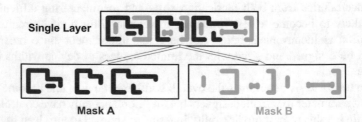

Fig. 6.3 Layout decomposition for double/multi-patterning: a single layer is decomposed in two or more masks to enhance the resolution

get the original intended layout. By doing so, the pitch size is effectively doubled thereby enhancing the effective resolution [22]. To achieve this, on the design side, coloring (mask) assignments are necessary. Examples of commonly used multi-patterning techniques include litho-etch-litho-etch double patterning (LELE DP), spacer-is-metal self-aligned double patterning (SIM SADP), litho-etch-litho-etch-litho-etch triple patterning (LELELE TP), and spacer-is-dielectric double patterning (SID SADP). To use these techniques designers can include the colored masks per layer that must be multi-patterned (with the support from EDA tools) or use a colorless flow then the foundry performs the decomposition [23]. In summary, multi-patterning will introduce additional complexity and considerations that need to be addressed during the design process.

6.2.3.2   Fin Formation

Although multi-patterning introduces new fabrication challenges, some of the known fabrication steps from the planar technology can still be repurposed to form new required shapes like the 3D fins. For example, sidewall spacer deposition steps from planar processes can be utilized to perform self-aligned double patterning (SADP). Similarly, the steps that were used to form shallow trench isolation (STI) can be extended to fabricate fins by additional etching of STI areas and thereby exposing Si fins. Fins are fabricated in regular parallel patterns over a large area, and unwanted fins are then "cut" (removed) with the remaining fins becoming part of active areas of the transistors; thus FinFET fabrication has been very compatible with the conventional CMOS processes by repurposing several existing procedures with a few extra necessary steps.

6.2.3.3   Fin Shapes

In Sect. 6.2.2, we have discussed how fin height affects digital design; several studies have shown that FinFET performance can also be affected by the cross-sectional area of the fin, which is determined by the fin shape. Intel's 22 nm node microprocessor was built with FinFET sidewalls sloping at about 8° from vertical which makes more sturdy devices among other advantages [25]. Figure 6.4 shows three mostly used types of fins from the literature. Experimental data showed that a FinFET with a rectangular cross-sectional area has better short-channel effect metrics, while in particular sub-threshold slope, GIDL and DIBL are less if compared with a triangular or trapezoidal cross-sectional area [24]. On the other hand, a triangular fin can reduce leakage current by 70% if compared with a rectangular fin [26]. Similar to fin height, although designers don't have control of the fin shape at design time, this can still be a codesign opportunity of close collaboration between foundry and design teams for picking the right fin shapes to achieve the optimal results from the technology.

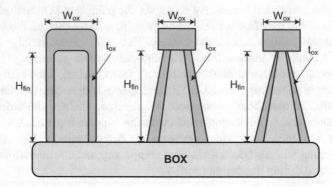

Fig. 6.4 Left side: a fin with a vertical slope which presents better short-channel metrics [24]. Middle: a standard fin with some degree of inclination as the one used in the 22 nm Intel's node [25]. Right side: a fin with a triangular cross-sectional area that can help to reduce the leakage [26]

6.2.3.4 Middle-End-of-Line (MEOL)

Middle-end-of-line (MEOL) is a new term introduced in the FinFET era—it refers to the intermediate process steps (contacts to gate and source/drain) that complete the transistor formation (front-end-of-line, FEOL) before interconnect and vias formation (back-end-of-line, BEOL) [27]. MEOL is necessary for providing better cell level connections with restricted patterning capabilities and multi-patterning [28]. MEOL increases the complexity of fabrication and modeling as well. For circuit designers, the added new parasitic effects from MEOL need to be addressed during the design process since these parasitics have been demonstrated being one of the dominant sources [29]. MEOL parasitics have been usually accounted at the logic gate-level parasitic extraction step using the standard EDA tools. For physical design engineers, the added MEOL means additional complex design rules and longer debugging process. From a tool and flow aspect, layout tools must automate conformance to rules as much as possible.

6.2.4 A Comprehensive Study of Bulk vs. FDSOI vs. FinFET Devices

From a digital circuit designer's perspective, whether the technology is planar or FinFET, whether it is bulk or SOI, the parameters of interest are quite similar—how much current can one transistor drive, PMOS to NMOS (PN) ratio, leakage, DIBL, GIDL, and so on. In this section, we present a study that is based on extensive simulations across multiple technology nodes. For the 7 nm node, we use a recently developed predictive 7 nm PDK [10] which is based on current realistic assumptions for the 7 nm technology node but is not tied to, or verified by, a specific foundry. We believe that this study gives good insights into how FinFET devices perform right now (with industry PDKs) and how good these devices are likely to be in the future (with predictive PDKs), compared to planar devices in older processes as we move forward. We summarize the results in Table 6.1, and details of each parameter are discussed in the following sections.

6.2.4.1 Device Models

Device models are critical for circuit designers to run simulations and study design trade-offs. Models need to be accurate and efficient in terms of simulation run time and complexity. The fact that fins are 3D structures that rise above the substrate means that they are more strongly affected by their immediate environment than planar devices. This brings a number of challenges for developing the device models. For example, the interactions between devices and their surroundings need to be accurately modeled. Besides, the unique gate structure leads to increased gate capacitance and also to more components when modeling the parasitic capacitances and resistance compared to planar devices [31, 32]. These capacitance and resistance values are very important since the inaccuracy caused during extracting R and C parasitics lead to mis-characterization and under/over-estimated design margins. Figure 6.5 shows an example of how FinFET parasitic capacitance is accounted for a 2-fin device. It is clear that more components contribute to both intrinsic capacitances (e.g., in the SPICE models) and extraction capacitances (e.g., accounted during extraction). For example, the gate capacitance includes gate to top of fin diffusion, gate to substrate between fins, gate to diffusion inside channel, gate to diffusion between fins, gate to contact, and more. Similarly, the fin-to-fin capacitance is also a new parasitic source introduced by FinFET devices. Therefore, the complexity of modeling has been increasing as the device dimensions shrink. Coupling and Miller effects become more pronounced in these devices as well.

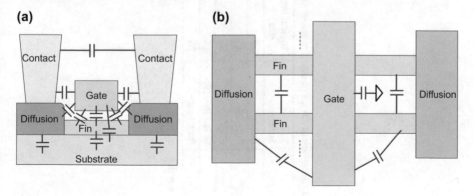

Fig. 6.5 Capacitance components for a FinFET device: (**a**) cross-sectional view and (**b**) top view

Table 6.1 Summary of device parameters across multiple technology nodes (most of these parameters are extracted based on simulated I–V curves)

Technology	Substrate	Physical length L_g (nm)	Nominal V_{dd} (V)	I_n/I_p (saturation)	Sub-threshold slope (mV/dec)	DIBL parameter	GIDL slope (mV/dec)	Channel length modulation λ (V)
130 nm	Bulk	120	1.2	4.24	92.07	0.53	3346	0.246
45 nm[30][a]	Bulk	45	1.0	1.45	98.3	1.61	286	0.387
28 nm	FDSOI	30	1.0	3.21	84.2	0.993	198.42	0.260
1 × nm	Bulk FinFET	14	0.8	0.99	71.1	0.485	429.79	0.256
7 nm[10][a]	Bulk FinFET	20	0.7	0.90	67.6	0.745	2220.6	0.203

[a]Predictive nodes

The structural differences between FinFETs and planar devices bring new modeling challenges. In a planar device, the source and drain are self-aligned with the gate and often intrude slightly under it. In FinFET devices there are spacers between the gate and the source and drain, which are usually raised and have a strain caused by a SiGe layer that creates a lattice mismatch. This means there are much more "hidden" parasitic capacitance and resistance structures, and more careful calibrations are required to achieve good accuracy. As for designers, the simulation efficiency also matters and it depends on the levels of model complexity, but thanks to the fast solvers and accurate extraction tools that have been developed recently, the simulation time has remained tractable.

6.2.4.2 Leakage

One of the driving forces that led the industry to move away from bulk planar to FD-SOI or FinFET technologies was the increase in leakage. With every new process generation the doubling of gate density is also associated with roughly a doubling of the amount of leakage current density [33]. This is also clearly shown from the simulation results in Fig. 6.6 where the sub-threshold current (OFF current) per unit width is plotted for different technology nodes. It can be concluded that, when scaling from (e.g.) 130 to 45 nm, the leakage current increases significantly, due to the fact that the channel underneath the gate becomes relatively deeper and a significant volume of the channel is too far away from the gate, thus there is a subsequent loss of electrostatic control of the gate. On the other hand, FDSOI and FinFETs have much less leakage because the gate has better control over the channel in these technologies. Our simulations, for example, show that 28 nm FDSOI and

Fig. 6.6 Leakage current evolution with technology scaling (*Predictive technology nodes (45 nm and 7 nm))

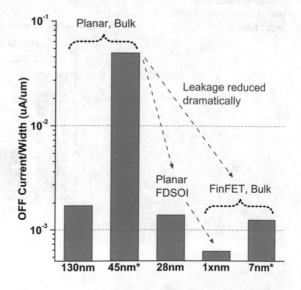

7 nm FinFETs have comparable leakage numbers, although 7 nm FinFET is five generations ahead. However, $1\times$ nm bulk FinFET shows an even higher reduction of leakage of at least 50%. This can be due to the fact that FDSOI and FinFET reduce leakage with slightly different mechanisms—in FDSOI, leakage reduction is achieved by making the channel thinner and by limiting its depth with the help of an insulating layer; while in FinFET it is achieved by wrapping the gate around the channel.

Another important parameter that is associated with leakage is the *sub-threshold slope*, which reflects how fast the device can switch from OFF to ON, with the lower bound being 60 mV/dec at room temperature. Table 6.1 shows that, due to the move to FDSOI and FinFET, the sub-threshold slope value has actually improved with scaling and this has resulted in a significant benefit for continuously improved performance, active power, leakage power, or a combination of the three over the past few years [34].

6.2.4.3 I_{on}/I_{off} Ratio

The I_{on}/I_{off} ratio is an important figure of merit for having high performance (higher I_{on}) and low leakage power (lower I_{off}) for transistors. Since the leakage current (I_{off}) has been significantly reduced in FinFET devices, their I_{on}/I_{off} ratio is superior to bulk, as shown in Fig. 6.7. This has also enabled a continuous performance improvement with scaling.

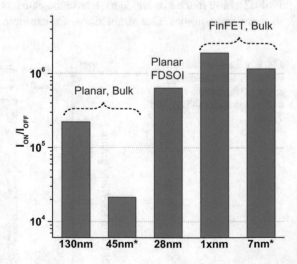

Fig. 6.7 I_{on}/I_{off} ratio evolution with technology scaling (*Predictive technology nodes (45 nm and 7 nm))

6.2.4.4 Drain-Induced-Barrier Lowering (DIBL)

Drain-induced-barrier lowering (DIBL) is a short-channel effect that appears as the distance between the source and drain decreases to the extent that they become electrostatically coupled. The drain bias affects the potential barrier to carrier flow at the source junction, resulting in sub-threshold current increase. To characterize it, we use the DIBL parameter, which corresponds to the change of leakage current due to V_{ds} and is defined in Eq. (6.1). Clearly small values for the DIBL parameter are preferred as they indicate less DIBL effect. Table 6.1 shows that FinFETs achieve very good DIBL behavior compared to bulk devices. In particular, the $1\times$ nm FinFET device has the lowest DIBL effect among all five technology nodes.

$$\Delta log(I_{\text{off}}) = (\text{DIBL Parameter}) \times V_{ds} \qquad (6.1)$$

6.2.4.5 Channel Length Modulation (CLM)

Channel length modulation (CLM) is another short-channel effect that is caused by large drain biases. It is characterized by the CLM parameter λ which is usually inversely proportional to the channel length; smaller λ reflects a lower CLM effect. Table 6.1 shows that CLM has been getting worse as the channel length shrinks in planar devices even with the increased doping density. When technology switched from planar to FDSOI and FinFET, CLM has been improved due to the better gate control over the channel. In particular, in the 7 nm technology node, the CLM effect is the weakest and is as good as in a relatively old long-channel 130 nm planar technology.

6.2.4.6 Gate-Induced Drain Leakage (GIDL)

The introduction of high-k/metal-gate stacks in planar devices has led to substantial reduction in the gate leakage but has exposed other leakage mechanisms such as gate-induced drain leakage (GIDL) as a primary gate-related leakage mechanisms [35]. GIDL occurs due to the high reverse bias between the silicon body and the drain junction (a PN-junction) near the gate edge at a near zero or a negative gate bias [36]. It usually increases as the gate length (L_g) decreases due to the floating body effect and is usually pronounced in short-channel devices. In this book, we pick the GIDL slope to quantify this effect; the larger this slope, the weaker the GIDL effect the device has. Interestingly, results shown in Table 6.1 indicate that as the technology switched to FinFET, GIDL has actually improved. The suppression of GIDL can be explained by the light doping of the channel and better junction placement gradient as suggested in [35]. In conclusion, our results so far confirm that FinFETs are superior to planar devices in terms of leakage, $I_{\text{on}}/I_{\text{off}}$, DIBL, CLM, GIDL, and thus appear to be a true "back to the future" reset of most of the metrics that were getting worse with every new technology generation for bulk planar technologies.

6.2.4.7 W_p / W_n Ratio

Another interesting observation for FinFET technologies is that the pull up network (PUN) and the pull down network (PDN[3]) can become truly symmetric. PMOS and NMOS devices with the same number of fins have very comparable driving strength, therefore the conventional 2:1 or 3:1 sizing strategy is not applicable (or necessary) any more in the FinFET case. This can be seen from the I_n / I_p ratio in Table 6.1, which is very close to 1 for both FinFET nodes. Figure 6.8 further demonstrates this. It plots the voltage transfer curve (VTC) under different supply voltages for a FinFET inverter sized with $W_p / W_n = 1$. It shows that the small-signal gain (which is the slope of the transfer curve when the input is equal to the mid-point voltage) is close to ideal (very high gain), and the curves are very balanced in all cases (even at very low voltages) which further demonstrates that the ratio of 1:1 is close to optimal for FinFET logic.

The reason behind this fact is due to the unique fabrication process for FinFET. As opposed to planar structures which can only be fabricated in a single plane due to

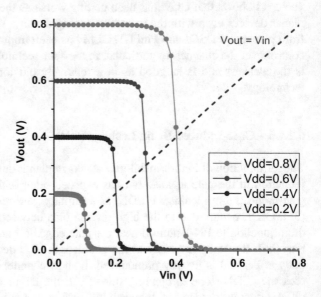

Fig. 6.8 VTC curves under different supply voltages for a 1× nm FinFET inverter (PMOS and NMOS are sized equally)

---

[3]Here PDN refers to the pull down network, in the rest of book, PDN stands for power delivery network.

process variation and interfaces traps, FinFETs can be fabricated with their channel along different directions on a single die—this results in enhanced hole mobility. The N type FinFETs implemented along plane $\langle 1\,0\,0 \rangle$ and the P type FinFETs fabricated along plane $\langle 1\,1\,0 \rangle$ lead to faster logic gates since it combats the inherent mobility difference between electrons and holes [3, 25, 37]. Moreover, since the gate has very good control over the channel, doping can be much lighter, thus allowing a reduction in random dopant fluctuations (RDF) [9] and mitigating the impact of mobility on current. The symmetry between PUN and PDN leads to ease in terms of physical design and sizing but it can also bring slight changes in design decisions and standard cell design.

6.2.4.8 Alpha-Power Law

The long-channel MOSFET model (i.e., Shockley model) assumes that carrier mobility is independent of the applied fields when the lateral or vertical electric fields were low [38]. However, for short-channel MOSFETs, the velocity of carriers reaches a maximum saturation speed due to carriers scattering off the silicon lattice; this also leads to a degradation in mobility that depends on the gate to source voltage V_{gs}.

The drain current is I_d quadratically dependent on the drain to source voltage (V_{ds}^2) in the long-channel regime and is linearly dependent on V_{ds} when transistors are fully velocity saturated due to an electric field higher than a critical electric field $E_c = V_c/L_g$ [39], where V_c is the corresponding critical voltage and L_g is the gate length. A moderate supply voltage is when the transistor operates between the long-channel regime and velocity saturation. The complete model, called the α-power law model, is presented in the following equations:

$$I_{ds} = \begin{cases} 0, & V_{gs} < V_{th} \quad \text{(Cutoff)} \\ I_{dsat}\frac{V_{ds}}{V_{dsat}}, & V_{ds} < V_{dsat} \quad \text{(Linear)} \\ I_{dsat}, & V_{ds} > V_{dsat} \quad \text{(Saturation)} \end{cases} \quad (6.2)$$

where $I_{dsat} = P_c \frac{\beta}{2}(V_{gs} - V_{th})^\alpha$ and $V_{dsat} = P_v(V_{gs} - V_{th})^{\alpha/2}$. The exponent α is called the velocity saturation index, and ranges from 1 for fully velocity saturated transistors to 2 for transistors with long channel or low supply voltage. We performed $I_{ds} - V_{gs}$ simulations for the base unit NMOS transistors of four different technologies and extracted their respective velocity saturation index α. The results are summarized in Fig. 6.9, which suggests that, as we switch to FinFETs, devices behave more similar to long-channel devices, thus FinFETs can again be considered a "back to the future" type of device.

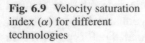
Fig. 6.9 Velocity saturation index (α) for different technologies

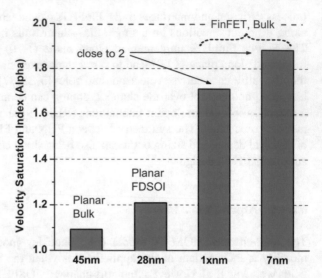

6.2.5   FinFET Circuits

Since FinFET devices have much better electrostatic properties and other metrics than planar devices, new logic and a larger design space become available. In this section, we discuss the new changes that FinFETs have brought from a circuit design perspective.

6.2.5.1   Logic Styles

As discussed in Sect. 6.2.2, FinFETs can come in two flavors—short-gated (SG) and independent-gated (IG). For IG FinFETs, because the top part of the gate is etched out, so this results in two independent gates that can be controlled separately, the four-terminal IG-mode FinFETs thus offer more design styles [11, 12]. Although the gates are electrically isolated, their electrostatics are highly coupled. The threshold voltage of either of the gates can be easily influenced by applying an appropriate voltage to the other gate. Shown in Fig. 6.10 is one example of different flavors of 2-input NAND gate implemented using SG/IG gate or a hybrid of both (adapted from [11]). In SG mode, FinFET gates are tied together, so they work similar to the planar devices. In IG mode, one device (with two gates) is driven by two independent signals, and some logic functions can be realized by one device; in IG-low power mode, one gate is disabled and acts as the reverse-biased back-gate. The designers can even mix the two types of devices and balance the trade-offs if this is allowed by the foundry, but it should be noted that IG gate requires one more step of etching in the fabrication step.

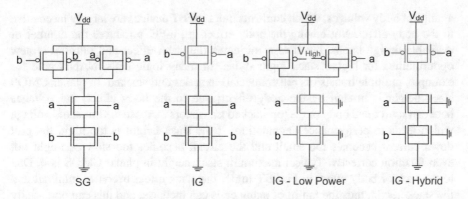

Fig. 6.10 Different FinFET logic styles: 2-input NAND gate designs with SG and IG devices

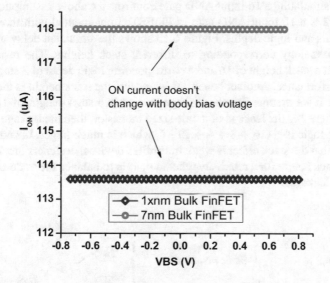

Fig. 6.11 ON current vs. body bias for a 2-finger $1\times$ nm and 7 nm NMOS transistor

6.2.5.2 Body Effect

Adaptive body biasing (ABB) has been widely used by circuit designers as an effective design technique to reduce the impact of die-to-die and within-die variations by changing the NMOS or PMOS threshold voltages independently in order to maximize performance [40]. FinFETs fabricated in bulk or SOI processes receive little benefit from body biasing since the body effect is very weak because the channel in the FinFET is mostly on top of the fin, which is away from the body. Thus the body bias techniques are not directly applicable to FinFET circuit design anymore [41]. To verify this, we apply both reverse and forward body bias to a 2-fin transistor and simulate the ON current for both $1\times$ nm and 7 nm nodes, with the results shown in Fig. 6.11. As expected, the ON current stays almost unchanged with a range of body voltages, which confirms that FinFET devices are largely insensitive to the body effect. Not having the body effect in FinFETs reduces the number of available design knobs, but it also mitigates the stack effect and introduces new opportunities for higher stack logic styles. In many logic cells, NAND gate for example, multiple transistors are connected in series and stacked. In planar CMOS stack height is limited by the body effect; due to the body effect, the voltages between source and body of the top stacked transistors increase the threshold voltage which leads to performance degradation; if the stack height is too high, the pull down current becomes too small and the circuit becomes too slow or might not even function correctly. Typical maximum stack height in planar CMOS is 4. Due to the lack of body effect in FinFET logic, the stack effect becomes minimal and the stack height, thus the fan-in of many cells can increase, and this can potentially shorten the logic depth and further reduce the delay and leaking paths. Our first attempt of simulating a 16-input AND gate confirms the above assumptions. Shown in Fig. 6.12 is a 16-input AND gate in FinFETs implemented with three different stack heights and logic depths. Figure 6.13 shows the simulated delay with $1 \times$ nm FinFET technology corresponding to different stack heights. The results clearly suggest that a stack height of 16 and a corresponding logic depth of 2 stages achieve the best performance. Another benefit of increasing the stack height is the reduction in leakage. If we assume that the leakage current in a stack-height-of-16 design is $\sim 16 \cdot I$, where I is the leakage of a unit-sized transistor, then the leakage for a stack height of 2 logic is $\sim(16 + 8 + 4 + 2) \cdot I$, which is much larger. In summary, due to the fact that the stack effect is weak in FinFET devices, designers are able to play with the stack height for a relatively relaxed margin to balance the trade-offs of area, delay, and leakage.

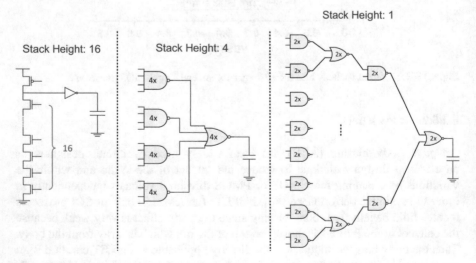

Fig. 6.12 A 16-input AND gate implemented with different stack height (1, 4, and 16)

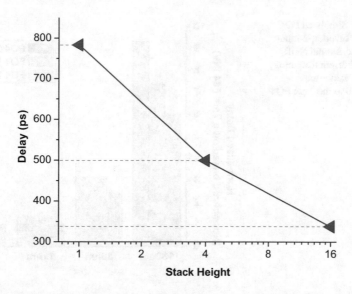

Fig. 6.13 16-Input AND delay simulations with different stack height (interconnect capacitance is considered)

Table 6.2 Normalized logical effort g and parasitic delay p values

	Textbook			130 nm bulk			28 nm FDSOI			1× nm FinFET			7 nm FinFET		
	INV	NAND	NOR	INV	NAND	NOR	INV	NAND	NOR	INV	NAND	NOR	INV	NAND	NOR
g	1.00	1.33	1.67	1.00	1.14	1.54	1.00	1.11	1.52	1.00	1.06	1.34	1.00	1.35	1.59
p	1.00	2.00	2.00	0.49	0.96	0.80	2.90	4.21	3.38	0.62	1.30	0.95	1.52	1.68	2.59

6.2.5.3   Logical Effort

The logical effort method is an approximate, simplified model to analyze the delay of a gate. The normalized delay is expressed as

$$d = f + p = g \cdot h + p \tag{6.3}$$

where p is the parasitic delay, i.e., the delay of the gate driving no external load, and f is the effort delay, expressed as the product of logical effort g and fanout h. The logical effort g is proportional to the complexity of a gate as a more complex gate leads to higher gate delay. The fanout h is the ratio of the output load capacitance to the input capacitance of a gate.

We estimate the g and p for an inverter, a 2-input NAND and a 2-input NOR across four technologies with simulations. The setup consists of *fanout of 1* and *fanout of 4* gate delay chains. The final results are summarized in Table 6.2. Values of g and p have been normalized to the respective inverter values for each technology. The table shows that the g and p values vary slightly across technologies because of the slight differences in transistor sizing strategies. Measured normalized delays for different

Fig. 6.14 Simulated FO4 delays for Inverter, 2-input NAND and 2-input NOR gates in different technology nodes (all values are normalized to the 7 nm FO4 INV delay)

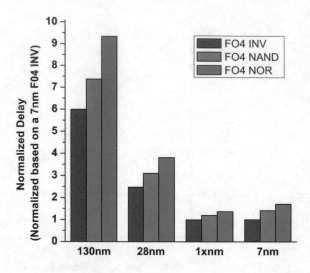

gates are presented in Fig. 6.14 which shows that gates maintain a similar trend for increase in complexity across different technologies. NOR gates with stacked PMOS are slower than NANDs stacked NMOS even in FinFETs where the ratio of ON current in NMOS to PMOS is close to 1 as listed in Table 6.1.

6.2.5.4   Standard Cell Libraries

There are many trade-offs to be considered when developing standard cell libraries. For example, logic offerings such as the maximum number of logical inputs on complex gates, flip-flop and latch offerings, clock buffers, and drive strength for each cell. As discussed in the previous sections, FinFET devices have several unique intrinsic device characteristics, and these bring several changes to the standard cell library designers as well. First, with planar transistors, designers can arbitrarily change transistor width in order to manage drive current. With FinFETs, due to the width quantization fact as discussed in Sect. 6.2.2, they can only add or subtract fins to size the device and change the current. Second, since body biasing is generally ineffective, as discussed in the last section, this might lead to more logical inputs on complex gates in FinFET libraries. When it comes to the physical design, the FinFET devices have periodic structures, and the optimal Wp/Wn ratio is almost 1:1, thus the FinFETs layout can be made more regular, and the PMOS and NMOS regions are symmetric. The standard cell template height (in the number of M1 or M2 wiring tracks) usually comes in several flavors. For example, a high density library might be 9 tracks tall, a high performance library might be 13 tracks tall, and a power optimized library might be 10.5 tracks tall. But in FinFETs, the additional constraint of fitting a fixed number of fins within a cell complicates this [6]. Additionally, in most FinFET technologies, fin and metal pitches are different and do not tend to line up. Power rail connections at the top and bottom of the cell typically force the removal of one fin each, and typically two additional fin tracks must be removed in the center of the cell to accommodate gate input connections. In addition, [6] pointed out that to meet the multi-patterning requirement, the coloring process needs to be conducted during the design of the standard cells, and coloring also needs to meet density solutions (each color mask must have reasonably consistent density across the chip). In summary, all of these make compact FinFET cell design more complex.

6.2.5.5   FinFET SRAM Design

SRAMs are one of the most area and power hungry components on chip. The never-ending demands for packing more functionality per area and delivering higher performance from processing units lead to continuous scaling of devices [42]. This scaling trickles down to smaller bitcells and enables an increase in memory array density in terms of number of bits stored per unit area. From a density point of view, minimum sized transistors are desired in bitcells which translates to a 1:1:1 (PU:PG:PD) fin bitcell for FinFETs (where PU is the size of the pull-up PMOS, PD is the size of the pull-down NMOS, and PG is the size of the pass-gate NMOS in a 6T SRAM cell). The 1:1:1 bitcell provides the highest array density but it suffers from flaws in terms of lower read stability and writability [42, 43]. The constant need to scale voltage for lower power further exacerbates SRAM readability and writability issues. This calls for alternate bitcells like the low voltage (LV) 1:1:2 cell and high performance (HP) 1:2:2 cell [42] along with read and write assist techniques to improve SRAM metrics. Several assist techniques [44, 45] have been proposed and studied to improve SRAM performance and lower operational V_{min}. These techniques focused on improving PD:PG strength ratio for read assists and PG:PU strength ratio for write assists. Such techniques become increasingly required for FinFET SRAM design because of the same width quantization effect which limits the device-level sizing options for improving SRAM bitcell functionality.

6.2.5.6   Thermal Effect Inversion (TEI)

Thermal behavior is one of the important device characteristics that affect the design decisions like margins, floorplan, and cooling systems. It has been shown in the literature that thermal characteristics of FinFET-based circuits are fundamentally different from those of conventional bulk CMOS circuits [46, 47]. In a bulk technology, if the transistor operates in the super-threshold region, the delay increases with the temperature, and in the near/sub-threshold region, the delay decreases with the increasing temperature. For FinFETs, it has been reported that the circuits actually run faster at higher temperatures in all supply voltage regimes (including the super-threshold region), and this is called the temperature effect inversion (TEI) phenomenon [46]. In both planar and FinFET devices, the threshold voltage decreases at a higher temperature, and the mobility of charge carriers in the channel decreases due to the ionized impurity and phonon scattering [48]. TEI happens due to the fact that FinFET channels are usually undoped or lightly doped, so they exhibit only a small change in mobility with temperature. It has been shown in [49] that TEI's inflection voltage approaches nominal supply and the impact of this effect can no longer be safely discounted when scaling into future FinFET and FDSOI devices with smaller feature sizes. To further study this, we simulated a 9-stage ring oscillator with multiple technology nodes and plotted the delay vs. temperature in Figs. 6.15 and 6.16; the results show that for all technologies, the increased temperature speed up the devices if they work under near and sub-threshold region. Interestingly, for the 28 nm FDSOI node, TEI appears across all voltages, and for $1\times$ nm bulk FinFET node, the TEI effect has already approached 0.7 V, which is only 0.1 V below the nominal voltage (0.8 V). Similarly, for 7 nm bulk FinFET, the inversion starts from around 0.6 V (0.1 V below the nominal voltage of 0.7 V). Based on this study, we can conclude that the TEI effect is indeed becoming increasingly important in current and future technologies as it will cover all of the operating voltage ranges.

The TEI effect can significantly affect the design trade-offs and cooling budgets. On one hand, a higher temperature increases the leakage and cooling budget, but, on the another hand, it helps with performance. The benefits of TEI can be maximized with the assist of novel power management techniques that dynamically tune the voltage or frequency based on the real-time temperature [49, 50] or novel algorithms that can determine the maximum performance under power constraints [51]. Since thermal issues also emerge as important reliability concerns such as wearout throughout the system lifetime, the TEI effect can compensate some of the performance degradation introduced by reliability threats such as BTI and EM [48, 52]. The optimal operating temperature can be exploited to reduce design cost and run-time operating power for overall cooling with the proper utilization of the TEI effect.

6.2.5.7 Variability and Reliability

Reduced feature sizes cause statistical fluctuations in nanoscale device parameters which are known as process variations. They lead to mismatched device behaviors and can degrade the yield of the entire wafer. In planar devices, a finite number of dopants are inserted in the channel which leads to random doping fluctuations (RDF) causing significant variations in threshold voltage. For FinFETs, since the channel is undoped or lightly doped, this reduces the statistical impact of RDF on V_{th}. The variability associated with line-edge roughness (LER), the random deviation of gate line edges from the intended ideal shape, which results in non-uniform channel lengths, is also lower in FinFETs. But process variations still impact FinFETs—

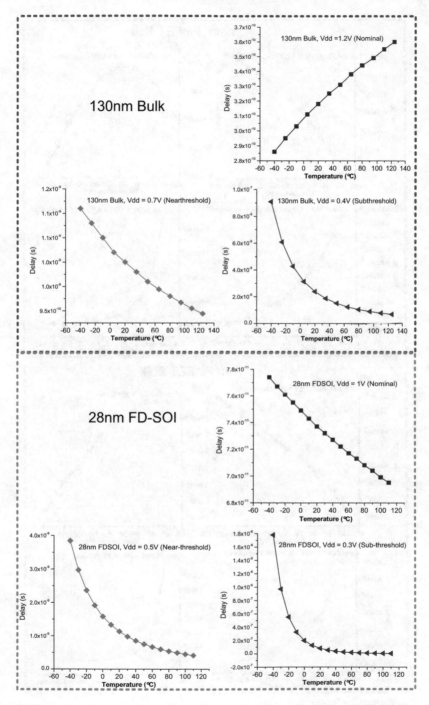

Fig. 6.15 Simulated thermal characteristics (delay vs. temperature) in planar technology nodes for a 9-stage ring oscillator. Navy blue—super-threshold; orange—near-threshold; red—sub-threshold

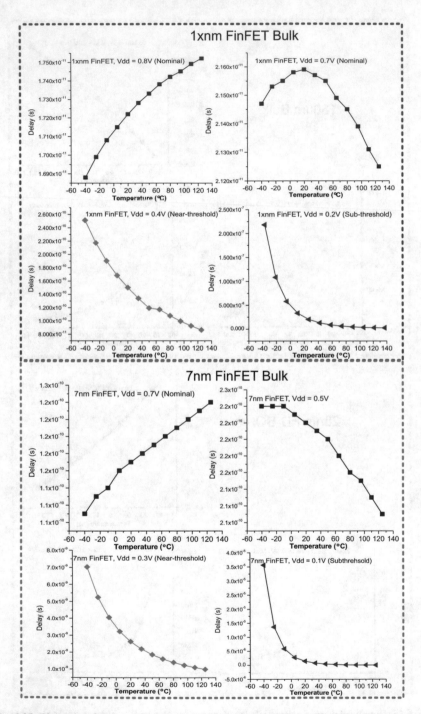

Fig. 6.16 Simulated thermal characteristics (delay vs. temperature) in FinFET technology nodes for a 9-stage ring oscillator. Navy blue—super-threshold; orange—near-threshold; red—sub-threshold

since they have small dimensions and lithographic limitations, they suffer physical fluctuations on gate length, fin thickness, or oxide thickness [3, 53, 54]. Overall FinFETs emerge as slightly superior to planar devices by overcoming RDF and LER, which are two major sources of process variations.

Besides process variations, which are considered "time-zero" process variability, time-dependent variations caused by wearout such as BTI and EM (which have been presented earlier in the book) result in critical variability and reliability considerations in FinFET devices as well. As discussed in Chap. 1, with scaling, on-chip components suffer more extreme voltage/current stress and higher current densities. Together these also lead to increased on-chip temperatures which potentially accelerate the wearout effects [55]. Besides, the thermal resistance (R_{th}) of the multi-gate topology and the reduced gate pitch in FinFET devices exacerbate self-heating which further accelerates wearout [56]. For EM, it can no longer be signed off using aggressive margins, instead a comprehensive thermal-aware EM signoff methodology needs to be adopted for FinFET designs. New types of EM rules that are dependent on the direction of current flow, metal topology, via types, and co-vertical metal overlaps are required to address the potential reliability issues [57]. A detailed study on FinFET wearout issues is presented in the second half of this chapter (Sect. 6.3).

6.2.5.8   Interconnect

As transistors themselves become smaller and smaller, the interconnect becomes more and more dominant in determining circuit performance. This is because the interconnect can't scale at the same rate as the transistors due to yield and EM requirements. As interconnect is becoming more compact at each node below 20 nm [58, 59], the interconnect RC parasitic delay affects the performance in a more significant way and can become one of the bottlenecks on the scaling roadmap. To address this, interconnect materials such as aluminum, cobalt (Co), or ruthenium (Ru) could be viable alternatives due to the better sheet resistance, but there are many cost and reliability considerations in the chosen interconnect scheme design [60]. The pitch size of the metal lines also doesn't scale down at the same rate as the technology when moving into the sub-20 nm regime due to the RC parasitic and coupling considerations. For designers, since they don't have control over the materials and design rules, the only knob they have is the dimension of the wire. This requires very careful considerations to address the interconnect capacitance even before the physical design. Many FinFET PDKs provide relatively accurate wire models that can be synthesized together with the designs in the early design phase. The FinFET PDKs usually support relatively accurate interconnect analysis to account for this.

6.2.6   FinFET Technology for Energy-Constrained IoT Applications

FinFETs provide improvements in power and energy consumption since they overcome the leakage problems of planar devices and deliver better performance. To further investigate this aspect, we simulate a NAND-based ring oscillator [61] across multiple technologies. The duty cycle of the ring oscillator is tunable—for this case we set it as 10%. Figure 6.17a plots the simulated delay vs. V_{dd}, in which the delay values of each node are normalized to the delay at their own nominal voltages. The simulations confirm that FinFETs provide a significant performance advantage under the whole ranges of operating voltages. Additionally, for FinFETs, the performance reduces much slower with the voltage scaling compared to planar

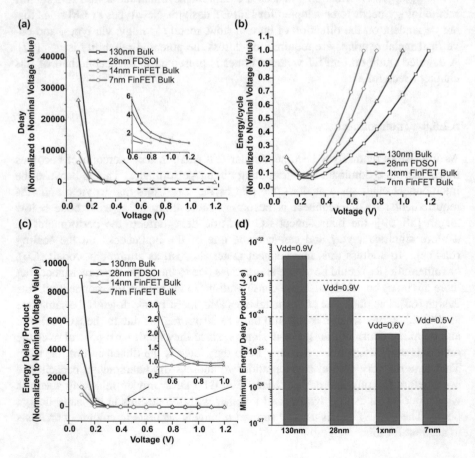

Fig. 6.17 (**a**) Delay vs. V_{dd}; (**b**) energy/cycle vs. V_{dd}; (**c**) energy delay product (EDP) vs. V_{dd}, and (**d**) minimum EDP values across multiple technology nodes (simulated with the same NAND-based ring oscillator structure)

technologies. Figure 6.17b plots the energy vs. V_{dd}, similar normalization strategy to (a) is applied. As can be seen, although the minimum energy optimal points are at similar voltages for all technologies (around 0.2–0.3 V range), the energy of FinFET scales the best with voltage; in other words, as voltage scales down, FinFETs offer more energy savings than planar devices. In Fig. 6.17c, the energy delay product vs. V_{dd} is plotted to show the energy efficiency. Since as the voltage scales down, the energy delay product doesn't change significantly for FinFETs compared to planar devices; thus FinFETs offer the best energy efficiency for circuit operating under a wide range of voltages. Figure 6.17d presents the minimum energy delay product (EDP) across the four technology nodes. As technology scales, the EDP improves as expected.

This study shows that FinFETs provide more options for leveraging trade-offs among performance and other metrics. For example, FinFETs offer very good energy efficiency across a wide range of voltages, voltage scaling techniques can be more powerful as IoT circuit designers strive to maximize performance per mW without hurting energy. FinFET-based designs are able to support a wider use of dynamic voltage frequency scaling (DVFS) and enable a wider range of applications that are not just limited to high-end performance-critical systems. By leveraging the costs, IoT applications can benefit significantly from FinFETs to achieve a combination of both optimal energy efficiency and great performance.

6.2.7 Summary: Digital Circuit Design with FinFETs

So far we have shown that FinFET devices offer significant performance improvements and power-energy reduction compared to planar devices. Digital circuit design with FinFETs broadens the design window once again—the operating voltage can still scale down, short-channel effects are reduced significantly, the process variations have been improved, there is lower leakage power in standby mode, and more.

Although FinFET devices offer advantages in many aspects, they also bring many challenges in the design process. We have studied some of them in the sections above, further summarized here. FinFET devices have non-standard shapes and require complex modeling of the parasitics in the TCAD tools. Moreover, the physical layout-dependent effects have a significant impact on several design metrics. Therefore, it is necessary for the design tools and flows to be able to assist designers to build circuits that accurately correlate to the models. During the design process, extraction plays a significant role in obtaining the accurate timing and power information, so necessary enhancements are required in the foundational EDA tools, in particular SPICE simulations, extraction, and physical verification that operate on the part of the design below the first metal layer [9]. As the interconnect resistance becomes dominant, IR drop and power-grid design becomes more critical. On top of these, standard cells, floorplan, and P&R need to be colored properly to meet the requirements of multi-patterning. For example, during power planning, all power rails need to be free of double patterning violations. Similarly, placement of all the standard cells and hard macros needs to be double patterning-compliant. Physical verification (e.g., DRC) engines need to be enhanced to check and guide the designers to meet the double patterning rules. More verifications are required, and more checkpoints need to be added during the design cycle to make sure the design specifications are met.

For custom designers and standard cell designers, all macros require a redesign due to the following reasons. First, the options of sizing are less granular due to the width quantization fact in FinFETs—getting more drive strength will require more fins in parallel. Second, options available for circuit designers are different than what they may be used to with planar devices. For example, body biasing is ineffective, thermal effect inversion (TEI) introduces new trade-offs, higher fan-in and complex logic are possible due to the insensitivity to the stack effect, etc. As dozens of complicated design rules arise for FinFET devices, physical design efforts are increasing, but the bright side for FinFET devices is the more regular layout and symmetric, roughly equal, P and N regions. Due to this, foundries usually provide a template layout for custom designers, on which fingers and gates are already laid out. It slightly eases the life of the physical designers, but the layout tools still need to automate conformance to rules as much as possible.

FinFETs offer more design options for trading performance with other metrics. As discussed in Sect. 6.2.6, one major benefit of FinFETs is the much higher performance with the same energy budget. Similarly, they consume much lower power and energy to achieve equal performance compared to planar devices. This essentially gives designers the ability to achieve the highest performance with the lowest power, which is a critical optimization for battery-powered devices. Since FinFETs have lower leakage and can operate faster, the circuit can afford to have more and fine-grained power gating structures to further save power in standby modes. Run-time techniques like DVFS can be used with a lower cost to the benefits. On top of these, FinFET circuits are able to operate in near-threshold to save energy with lower performance penalties [62].

As more transistors fit on a single chip in the FinFET era, design flows need to be able to handle very large designs which have billions of transistor at the full-chip level, thus optimizing the run time and reducing peak memory of the tool become increasingly important. Adding parallel execution within a run is also required. Because of the increased complexity and number of instances on chip, an increasing number of signoff corners are required to cover process and environmental variations. Addressing these new challenges together with the new, and more complex design-for-manufacturing rules, including double/multi-patterning, along with the increasing design scale, requires very close collaboration among foundries, tool vendors and designers to take full advantage of what FinFETs have to offer.

6.3 When "Things" Get Older: Exploring Transistor Aging in IoT Applications

6.3.1 Motivations

In the first half of this chapter we considered the opportunities and challenges introduced by FinFET technologies. Due to the high energy efficiency and level of integration, FinFETs are going to play an important role in many energy-constrained applications such as Internet of Things (IoT) [63, 64]. As the number of connected devices and connections between human and "things" increase rapidly, Internet of Everything (IoE) emerges as a wider concept of connectivity platform, where IoTs are key components and lay the foundations for the massive interactions with the world [65, 66]. IoT is a general-purpose technology by nature and can range from health care, transportation to agriculture and almost all aspects of life [67]. These applications impose common requirements for IoT devices such that they should have small form factors in terms of physical dimensions and weight. Since many IoT devices are battery-powered or batteryless, they require high energy efficiency and extreme low power consumption. In addition to these characteristics, IoT devices need to withstand hostile environments such as increased and highly variable temperatures and voltage noise [67]. More importantly, IoT nodes are required to operate reliably for a long time (e.g., decades), which translates to strict reliability requirements that can be affected by device degradation-induced circuit aging.[4] As we discussed in detail in the previous chapters, on-chip elements such as transistors and metal wires age gradually when under stress, and this can lead to permanent failures. Many IoT applications (like automotive or implantable medical devices) require almost zero error during the whole lifetime [68]. Harsh environments, such as high temperature, accelerate aging. As the number of on-chip elements scales up, more transistors are susceptible to aging and this leads to an increase in the system failure rate. Advanced technologies such as the FinFETs discussed in the previous section impose more aging issues than previous generations due to self-heating, reduced oxide thickness, narrower metal, and increased current density [69].

As IoT is becoming a broad concept, and circuit aging is a threat to the lifetime of IoT devices, it is necessary to understand how various future IoT systems are impacted by aging and what are the right solutions within the IoT context. In this section, we present a study of these concerns using extensive circuit-level simulations with foundry-calibrated aging models in an advanced $1\times$ FinFET node. As aging is highly dependent on application behaviors that eventually define the operating voltage, temperature, and active time, we first perform a survey of existing IoT applications and classify them based on aging-related metrics. The aging behaviors under each category are presented in the following sections.

---

[4]This book focuses on circuit aging. Battery aging and socket (and holder solder) aging are out of the scope of discussion.

6.3.2   State-of-the-Art Understanding of Circuit Aging for IoT

As has been detailed in Chaps. 2 and 3, the primary cause of circuit aging is the electrical stress affecting the on chip components like transistor, dielectrics, and interconnects. In general, there are mainly three dominant causes of aging in semiconductor devices. Bias temperature instability (BTI) and hot carrier injection (HCI) lead to transistor degradation [70], and electromigration (EM) increases the metal wire resistance. The main mechanisms of BTI and EM have been discussed in the previous chapters. HCI is another transistor aging mechanism that shares many similarities to BTI, both of them impact transistor parameters (e.g., threshold voltage V_{th} and carrier mobility μ) at a level that depends on the operating conditions and the usage of the circuit. As illustrated in Fig. 6.18, BTI is mainly caused by constant electric fields that degrade the gate dielectric. HCI is also caused by electrical field, but it happens mainly on the drain side and primarily occurs during switching. While BTI is partially reversible HCI is mostly irreversible. The parameter shifts of both effects are highly temperature dependent since temperature affects the interface trap generation. In this chapter, we mainly focus on transistor aging (BTI and HCI). EM is less impacted for IoT because the current density for IoT applications is relatively low.

Transistor aging has been studied for a long time mainly in aerospace and safety applications but it didn't gain interest in consumer devices until very recently. For example, in automotive or industrial IoT applications aging happens even when the system is inactive most of the lifetime due to the continuous constant stress across a significant number of transistors. Besides, many devices need to function under all possible conditions during their expected lifetime [71]. While a car today may sit idle 90–95% of the time, an autonomous vehicle might only be idle 5–10% of the time [72]. For another example, some medical implants require a reliable operation for more than 50 years [73]. In the previous studies of IoT, most of the attention has been paid on battery or package aging [74] while just a few studies have looked into circuit aging. For example, [75] introduced a method to obtain multi-threaded switching activity signatures for aging analysis in IoT applications but the focus was on the architectural level framework. Similarly, [76] proposed a solution that leverages the workload dependent reliability analysis for early product failure rate calculations for automotive applications. [73] provided a device-level

Fig. 6.18 Transistor aging: HCI occurs mainly during switching; PBTI happens when NMOS is under stress; NBTI happens when PMOS is under stress. BTI aging partially recovers during OFF states

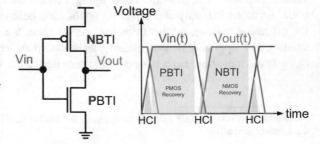

experimental study of BTI aging in ultra-low power applications. [77] proposed a unified model which captures the joint impact of random telegraph noise (RTN), BTI, and process variations (PV) within a probabilistic reliability estimation for near-threshold circuits. Most of the previous studies focused on circuit aging only for a very specific application or framework. Also previous works used analytical aging model which can lead to inaccurate or unrealistic predictions. The main goals of our study are as follows:

- We study aging impact with foundry-calibrated model instead of predictive models, by doing this, we provide more realistic predictions;
- We investigate transistor aging on a wide spectrum of IoT applications and provide a deep and realistic understanding of how aging affects each IoT category;
- We present several potential design solutions that address aging in IoT specifically, including ones that are inspired by the recovery techniques we have discussed throughout this book.

6.3.3   IoT Application Domains

The hardware requirements for IoT devices are determined by how and where they are deployed [67]. To develop a quantitative understanding of these requirements for IoT domains, we first survey the published SoCs and commercially available IoT products, covering applications such as agriculture/environmental sensors, automotive, industrial processes to medical implantables, smart cities, and consumer electronics.

We classify the existing IoT applications into ten groups mainly based on the usage and scale of users [67, 78] and summarize them in Table 6.3. We mainly extract the aging-related metrics, i.e., voltage, temperature, lifetime requirements, and active-time which is how long transistors are under stress. IoT chips in all categories are found to operate under super-threshold voltages during active phases of computation for performance purpose even in battery-operated systems. Most of commercial low-power IoT systems improve energy efficiency through heavily optimized deep sleep modes or minimization of the unnecessary on-chip components [67]. There are many ongoing attempts to operate IoT systems completely in near/sub-threshold regime to achieve significant energy efficiency improvements, especially in applications such as medical devices, sensors, and wearables, where energy harvesters can be adopted (applications 1, 2, 7 in the table). But this comes at the expense of performance and increased sensitivity to variations. In this work, we mainly look into IoT chips that operate at nominal voltage. Details of each application domain are explained in the following:

- Applications 1 and 2 represent personal IoTs where implantable devices usually operate continuously at human body temperature while consumer electronics such as wearables are exposed to environment. Implantables are active most of

Table 6.3 Summary of IoT applications specifications (aging-related metrics)

Applications	Temperature[a]			Core voltage[b]			Lifetime requirements[c]			Active time[d]		
	L	M	H	L	M	H	L	M	H	L	M	H
1—Implantable/heathcare		★		✓	✓	★			★			★
2—Consumer electronics/wearables	★	★		✓	✓	★	★				★	
3—Automotive			★			★			★		★	
4—Industrial processes	★	★	★			★		★				★
5—Public transportation	★	★				★		★			★	
6—Energy management	★	★	★			★		★			★	
7—Smart homes/buildings/cities	★	★		✓	✓	★		★				★
8—Retailing/malls		★				★		★			★	
9—Agriculture/environmental	★	★				★	★				★	
10—Wildlife/nature preservation	★	★				★			★		★	

★: Specification for current commercial products (in market and literature)
✓: Specification for future products (still ongoing research)
[a]For temperature, L—low ($\leq$27 °C), M—medium (27–100 °C), H—high (>100 °C)
[b]For voltage, L—low (sub-threshold), M—medium (near-threshold), H—high (nominal voltage)
[c]For lifetime requirements, L—low ($\leq$3 years), M—medium (3–10 years), H—high (>10 years)
[d]Here, "Active" means transistors are under stress. In many IoT applications, even when the systems are in "sleep" mode, many circuit blocks (such as accelerometers in a wristband) are still active and under aging stress. L—low (active time <20% of the lifetime), M—medium (active time 20–80% of the lifetime), H—high (active time >80% of the lifetime)

the time and require a relatively long lifetime—almost human being lifetime. On the other hand, wearables have a relatively shorter life cycles (around 3 years) and are inactive most of the lifetime.

- Applications 3 and 4 are industrial IoTs in which automotive sensors monitor the state of the vehicle and mostly reside inside engines. Thus they operate under a very high temperature and require a reliable operation throughout car's life span of more than 10 years. Similar monitoring strategies are used in industrial environment such as storage warehouse or product lines.
- IoT devices also enable ubiquitous sensing in city and home infrastructures (applications 5–8) where sensors are installed both indoors and outdoors. Thus they experience room or environmental temperatures and expect longer lifetime since frequent checking and repairs are impractical to these systems.
- Applications 9 and 10 represent environmental IoT applications. They have similar temperature requirement as city/home-scale IoTs (applications 7 and 8). The agriculture sensors usually last for one cycle of crops but other environmental IoTs require to last longer because they are distributed at a very large scale and many of them are not quite accessible physically once they are installed.

6.3.4 Simulation Results

As discussed in Chap. 1, process, voltage, and temperature (PVT) variations require additional timing margins that stretch the clock cycle. Similarly, guardbands are also required for aging. In this section, we present a study on the impact of aging by using the foundry-calibrated models in a $1\times$ FinFET technology with the Cadence reliability simulator RelXpert (integrated within Virtuoso ADE). Both BTI (including recovery) and HCI mechanisms are captured in the model.

6.3.4.1 Single Transistor Aging Under Different Conditions

Transistor aging causes parameter shifts, such as increased threshold voltage V_{th} and reduced mobility μ and this leads to a reduced ON current I_d. Two sets of simulations (DC and AC stress) are run at different temperature and lifetime conditions for single transistor under nominal voltage; the ON current degradation (%) is plotted. Figure 6.19 shows the DC stress case, in which the PMOS transistor ages continuously during the whole lifetime without any recovery—this gives the worst-case estimation of NBTI aging. Figure 6.20 shows the results where the transistor is under AC stress with 50% duty cycle allowing BTI recovery following stress. This includes both BTI and HCI degradations and provides an average case aging estimation. To account for how these degradation percentages compare to other variation sources such as process variations, we also run Monte Carlo (MC) simulations which give an estimated margin for process variations. As a reference, MC simulations show that σ/μ for ON current is around 7%. The aging-induced degradations are comparable to this, and at high temperatures, they can be even more

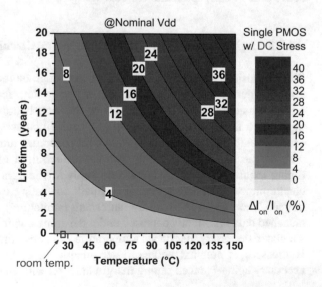

Fig. 6.19 Single transistor ON current degradation due to aging under *DC nominal* voltage stress

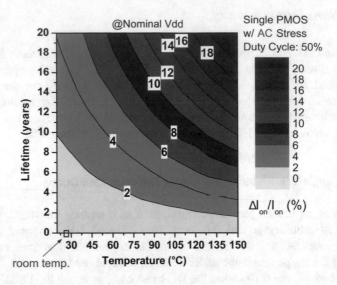

Fig. 6.20 Single transistor ON current degradation due to aging under *AC nominal* voltage stress with 50% duty cycle. Degradation is about half of the *DC* stress case due to recovery

than 10%. This comparison shows that under certain conditions, aging-induced margins need to be close or even larger than variation-caused margins. Another observation from Figs. 6.19 and 6.20 is that AC stress-induced degradation is almost half of the DC case and indicates that the degradation under the same temperature and lifetime condition can be roughly estimated as linearly proportional to the stress time. This assumption will be used in the following analysis for estimating how the active time of each application affects aging.

6.3.4.2 Aging-Induced Timing Failures in IoT Circuits

Single transistor aging causes an ON current degradation which can cause timing errors at the circuit level and failures at the system level. Figure 6.21 shows a typical datapath from the output (Q) of the launch flop to the data input (D) of the capture flop. Aging slows down each unit (increases $t_{datapath}$ and t_{setup}) thus causing setup time violations. As this effect becomes more significant in datapaths with deeper logic, extra timing margins are required to be added to meet setup timing requirements. To quantitatively study this margin under different operating conditions, we simulate a similar datapath consisting of a chain of inverters and buffers. The margin is found by increasing (stretching) the clock period until the launched data is correctly captured under the aging conditions. Figure 6.22 plots the simulated necessary margin vs. temperature and active time (how long the transistor is stressed) at nominal voltage. Each margin value has been normalized to the necessary aging-induced timing margin at 27 °C with an active time of 2 years as a

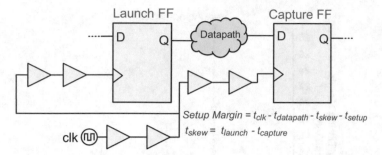

Fig. 6.21 Simulation setup (an example of datapath): aging can lead to timing failures such as setup violation by slowing down the datapath. Designers should take extra margins based on aging impact

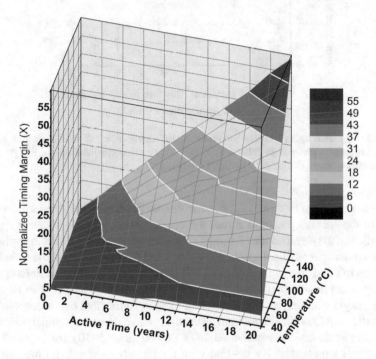

Fig. 6.22 Normalized timing margin vs. temperature and active time: margin shown on Y-axis is normalized to the required aging margin for datapath (shown in Fig. 6.21) for 2 years at room temp (27 °C)

baseline case for comparison. This baseline margin is equal to the required margin for temperature varying from 27 to 110 °C at time zero.

Based on the simulated results shown in Fig. 6.22, we apply the operating conditions of different categories of IoT applications summarized in Table 6.3 and list the estimated aging margin for each category in Fig. 6.23. As shown in the table, even for the same application category, different IoT devices operate at different

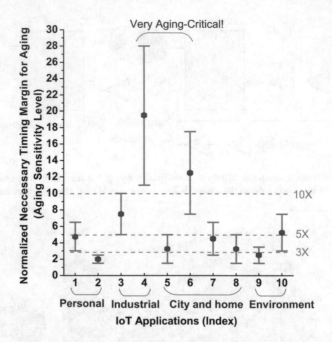

Fig. 6.23 Estimated aging margins for different IoT applications: X-axis corresponds to IoT application indexes grouped in Table 6.3, Y-axis shows the normalized design margin and the error bars show design margin range within each category

temperatures and active time. The error bar in the figure provides an estimated range of the necessary margin across the use scenarios within one category, and it also indicates the significance of aging (estimated aging levels) for each application. The two most aging-critical applications are 4 and 6, which correspond to industrial processes IoT and energy management IoT, respectively, in which high temperature and long active time are expected. These applications impose a more than $10\times$ larger design margin with respect to the baseline margin. In the next tier, automotive IoTs (application 3) also suffer significant aging due to high operating temperatures. Most of the city scale and environmental IoTs (applications 5–10) are exposed to the environment temperature but operate with a relatively long active time and hence they lie in a third tier ($3\times$ to $5\times$). Similarly, implantable devices operate at body temperature; however, they need to operate reliably for a long life span, so they are also on the third most critical aging level. As consumer electronics or wearables (application 2) are usually updated within a timescale of a couple of years, they are the least aging-critical, but even so, they still need to be margined properly to guarantee reliable operation during their lifetime.

6.3.4.3 Impact of Aging on IoT SRAMs

SRAMs act as an external cache for many of today's IoT applications [79]. They usually occupy the largest area of a SoC and may interact with multiple cores and functional units, hence it becomes imperative to study the impact of aging on SRAMs because the access-time and drive strength degradation may lead to timing failures across the chip. The access-time is directly proportional to the SRAM read current, I_{read}. Figure 6.24 shows the I_{read} degradation of a 6T SRAM across different temperatures for different active time. Each category of IoT applications is mapped on the plot. The I_{read} values have been normalized to time-zero I_{read} at 27 °C. The figure shows that many IoT applications incur more than 5% degradation during their lifetime with some critical applications such as industrial IoT experiencing more than 20% degradation in read current. Such a huge loss in SRAM performance caused by transistor aging may lead to fatal timing errors and hence should be addressed during the memory design cycle. The design process should also appropriately assign timing margins for aging based on target applications and their behaviors.

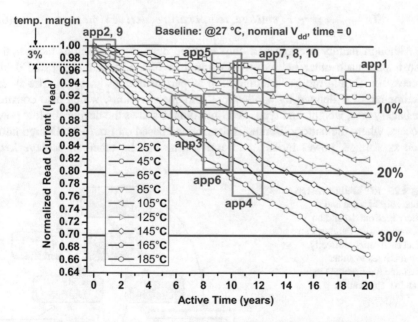

Fig. 6.24 6T SRAM reads current degradation with aging for different temperatures (nominal voltage)

6.3.5 IoT Lifetime: Battery vs. Chip Lifetime

In many IoT application, battery replacement is impractical due to the larger numbers of devices and inaccessibility of each node; a foremost requirement is thus that they can't rely on battery replacement. Thus a common way of defining lifetime of a battery-powered IoT system is by battery lifetime which is the time a node will operate in its normal mode without the need to replace the battery [67]. This is given by

$$T_{lifetime|battery} \sim E_{battery}/P_{average} \tag{6.4}$$

As transistor aging may lead to chip failure that might not be recoverable, as studied in the previous section, we propose here that the aging-induced chip lifetime should also be incorporated to determine the overall IoT system lifetime, which thus becomes:

$$T_{lifetime|Final} = min\{T_{lifetime|battery}, T_{lifetime|chip}\} \tag{6.5}$$

$$T_{lifetime|chip} \sim F(voltage, temperature, active\ time) \tag{6.6}$$

Although battery lifetime and chip lifetime depend on different factors, they also impact each other indirectly. Illustrated in Fig. 6.25 is a suggested design process for closing the IoT lifetime loop as part of the design cycle. The system lifetime target is defined by applications and specifications, which also constrain the battery size, weight, and type. The right branch shows the design-for-low power process, where the battery lifetime is determined based on Eq. (6.4). Design knobs such as voltage, power modes, and active time can be tuned to achieve lower

Fig. 6.25 IoT lifetime design cycle: chip lifetime and battery lifetime depend on different factors, but they can affect each other indirectly. Two lifetimes together determine the lifetime target of an IoT application

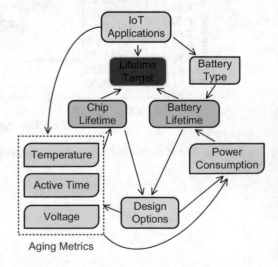

power consumption while fulfilling the performance requirements. Many of these design knobs are also constrained by the application itself, e.g., implantable devices (application 1) need to be active continuously and require fast response. On the left branch of the IoT lifetime loop in Fig. 6.25, the chip lifetime affected by aging is also highly dependent on knobs such as voltage, temperature, and active time. To guarantee that the final IoT system meets the lifetime target, the chip lifetime $T_{lifetime|chip}$ constrained by aging needs to be longer or equal to the battery lifetime $T_{lifetime|battery}$. The design margin needs to be assigned appropriately, but, as shown in the previous section, this margin can be very large in many IoT applications and can translate into wasted energy in the early lifetime, which in turn shortens the battery lifetime. As discussed in Chaps. 1 and 4, as an alternative, aging can also be addressed by adaptive solutions or recovery solutions to reduce the necessary design margin but at the expense of additional overhead from tracking or recovering. IoT systems are unique in many ways, thus new methods need to be explored to leverage the available trade-offs to meet the expected overall lifetime, careful design decisions considering both chip and battery lifetime being required. In the following section, we discuss several of such candidates.

6.3.6 Potential Design Solutions for Addressing IoT Circuit Aging

Adding design margins is currently the most common way of addressing aging in the design flow. This is a static solution where all transistors are margin-degraded by a certain amount based on the operating conditions. The difference in performance of an aged cell versus the original cell is computed and the ratio (aged/fresh) is used to derate cells in the design. But the large margin in many applications (shown in Sect. 6.3.4.2) is likely over-conservative and introduces a large performance penalty in the early lifetime. An alternative solution is to either recover aging using the on-chip solutions this book has addressed in Chaps. 4 and 5 or adapt to aging dynamically so that the design margin requirement can be relaxed. This section briefly discusses the feasibility of applying several such solutions for IoT.

6.3.6.1 Lowering the Operating Voltage

Operating the circuit in near/sub-threshold has been shown to be a very effective way of reducing the energy per computation and extending the battery lifetime especially in health care and body sensor IoT applications [67]. Meanwhile, aging has a power-law dependence on stress voltage [70], thus operating at lower voltages can suppress aging significantly. However, the challenges are performance loss and increased sensitivity with respect to other variations. The easy solution for performance degradation is to raise the operating voltage as necessary to meet the speed requirement and mitigate the impact of process variations, but this in turn will accelerate aging. One alternate method is to have fine-grain voltage domains which can ensure that voltage boosting is kept small enough so that aging does not introduce large degradations and the impact is locally constrained to critical sub-blocks. Fine-grain voltage domain can also maximize the opportunities to correct variations in paths that are critical due to aging and other variations. But this approach certainly leads to significant area overhead and design effort. Such trade-offs need to be addressed based on the actual available budgets that are defined by IoT applications.

6.3.6.2 IoT Circadian Rhythms: Active Recovery During Standby

Another widely used technique for energy savings in IoT circuit design is to put the circuit in "standby" state for as long as possible. This is feasible since these devices typically don't need to be active all the time in many applications as shown in Table 6.3. As discussed in the previous chapters, aging mechanisms such as BTI are recoverable when the transistor is OFF (not under stress) [70]. Hence the existing standby periods can be also utilized for recovery. Figure 6.26 (top) illustrates the power and aging profile of a typical IoT node, where the sensing activity is usually periodic and can be triggered by some real-time event. One solution to save power while reducing aging is to operate the whole processing unit at the lowest voltage level while maintaining its state in retention mode. Although lowering voltage can lead to some level of recovery, the transistors are still under stress (at a relatively lower level), and aging eventually gets triggered. An alternative solution is to turn off certain blocks completely through power gating and save states in retention registers. This will be essentially the *passive recovery* case in Chap. 2, where recovery occurs very slowly and incompletely [80]. The third option is to use the wearout-aware power gating structure described in Sect. 4.2.3 to reverse gate bias on the transistor and heal it faster; this active recovery approach enables maximum recovery during the IoT standby mode. Although effective for dealing with aging, the last two approaches come with extra power and area costs from power switches, logic retention, signal isolation, and additional floorplanning constraints [67]. The third solution also introduces one more voltage source and domain (the overhead is listed in Table 4.2), but for extremely aging-critical applications such as application 3, 4, and 6 shown in Fig. 6.23, necessary recovery techniques with acceptable overhead need to be employed to prevent system failures as reliability for these systems is one of the topmost concerns for designers.

6.3.6.3 Dynamic Margins Across Multiple IoT Applications

To minimize design effort and costs, circuit designers and chip vendors usually aim to use one SoC platform across multiple IoT applications. As even within one IoT application category, the operating conditions may change over time, these

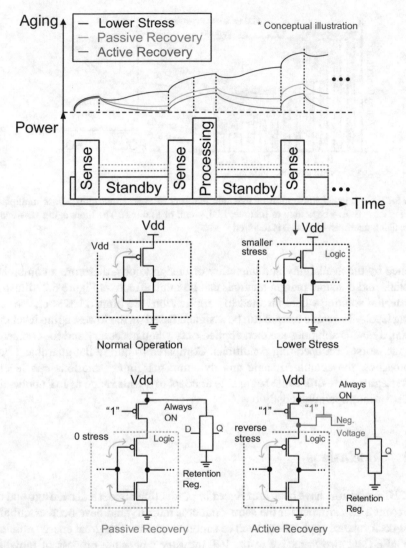

Fig. 6.26 Top—power and aging profile of a typical IoT node: this figure is for illustration only, the height and width are conceptually marked. For aging profile, Y-axis "aging" corresponds to aging-induced metric change such as ΔV_{th} or timing margin (reduced performance). Bottom—four different possible operating scenarios for IoT circuit, the active recovery technique can be applied to fully take advantage of the IoT intrinsic sleep behaviors

variations necessitate run-time compensation of aging such as techniques proposed in [81], where aging events (e.g., delay change) are tracked during operation. Once the failure flag (e.g., timing failure) is triggered, adaptive solutions such as dynamic voltage and frequency scaling (DVFS) or error correction are employed to compensate for the degradation. Unfortunately pure dynamic solutions can be

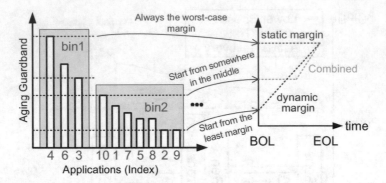

Fig. 6.27 Conceptual illustration of dynamic margins to enable one chip across multiple IoT applications (BOL—beginning of lifetime, EOL—end of lifetime). The more aging sensitive, the large aging guardband (Y-axis) is required

limited by the availability and tunability of design knobs, therefore a combination of static and dynamic margin methods can be more effective. Figure 6.27 illustrates a potential solution where targeted IoT applications are binned based on estimated aging levels. The static margin can be assigned based on the lowest aging level in the group, dynamic solutions are then applied only when necessary and for compensating the worst-case operating conditions. Compared to purely flat guardband-based approaches, the combined static and dynamic margining solutions can leverage power/aging trade-offs while being able to adapt to a wide range of IoT applications under various operating conditions.

6.4   Conclusions

FinFET transistors have been introduced in production almost a decade ago and they represent a new frontier for the semiconductor industry and have been essential for high-performance applications such as supercomputers. The great energy efficiency of FinFETs is also attractive to the IoT industry where the process of moving to FinFETs is still in transition. As highlighted in Fig. 6.28, in this chapter, we first studied the changes since the advent of the FinFET devices and addressed the challenges designers face with these new devices. FinFETs offer benefits in many aspects, such as significant improvements in power and performance metrics and improved short-channel behaviors. FinFETs offer advantages of extreme scaling while offsetting the limitations introduced by scaling for planar CMOS. However, design challenges also appear due to many of the unique properties of FinFETs. The first part of the chapter aims to add to the growing FinFET design knowledge base.

IoT applications and their diverse markets lead to a plethora of new requirements for IoT reliability. In the second part of the chapter, we presented a study that showed the FinFET transistor aging introduces new challenges for several IoT application

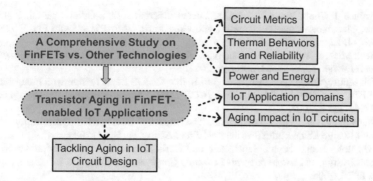

Fig. 6.28 Chapter 6 highlights

domains. The study also demonstrated that transistor aging should be taken into consideration earlier in the system design cycle. We also previewed a set of static and dynamic solutions (e.g., active recovery) to compensate or fix aging in IoT systems by fully taking advantage of the IoT intrinsic sleep periods.

References

1. Gordon E Moore. Cramming more components onto integrated circuits. *Proceedings of the IEEE*, 86(1):82–85, 1998.
2. Paolo Madoglio, Hongtao Xu, Kailash Chandrashekar, Luis Cuellar, Muhammad Faisal, William Yee Li, Hyung Seok Kim, Khoa Minh Nguyen, Yulin Tan, Brent Carlton, et al. A 2.4 GHz WLAN digital polar transmitter with synthesized digital-to-time converter in 14nm trigate/FinFET technology for IoT and wearable applications. In *Solid-State Circuits Conference (ISSCC), 2017 IEEE International*, pages 226–227. IEEE, 2017.
3. Debajit Bhattacharya and Niraj K Jha. FinFETs: From Devices to Architectures. *Advances in Electronics*, 2014, 2014.
4. Benton H Calhoun, Yu Cao, Xin Li, Ken Mai, Lawrence T Pileggi, Rob A Rutenbar, and Kenneth L Shepard. Digital Circuit Design Challenges and Opportunities in the Era of Nanoscale CMOS. *Proceedings of the IEEE*, 96(2):343–365, 2008.
5. James Warnock. Circuit Design Challenges at the 14nm Technology Node. In *Proceedings of the 48th Design Automation Conference*, pages 464–467. ACM, 2011.
6. Robert Aitken, Greg Yeric, Brian Cline, Saurabh Sinha, Lucian Shifren, Imran Iqbal, and Vikas Chandra. Physical Design and FinFETs. In *Proceedings of the 2014 on International symposium on physical design*, pages 65–68. ACM, 2014.
7. Jong-Ho Lee. Bulk FinFETs: Design at 14 nm Node and Key Characteristics. In *Nano Devices and Circuit Techniques for Low-Energy Applications and Energy Harvesting*, pages 33–64. Springer, 2016.
8. Bin Yu, Leland Chang, Shibly Ahmed, Haihong Wang, Scott Bell, Chih-Yuh Yang, Cyrus Tabery, Chau Ho, Qi Xiang, Tsu-Jae King, et al. FinFET Scaling to 10 nm Gate Length. In *Electron Devices Meeting, 2002. IEDM'02. International*, pages 251–254. IEEE, 2002.
9. Jamil Kawa. Designing with FinFETs: The Opportunities and the Challenges. In *Synopsys White Paper*, pages 1–8. Synopsys, 2012.

10. Lawrence T Clark, Vinay Vashishtha, Lucian Shifren, Aditya Gujja, Saurabh Sinha, Brian Cline, Chandarasekaran Ramamurthy, and Greg Yeric. ASAP7: A 7-nm FinFET Predictive Process Design Kit. *Microelectronics Journal*, 53:105–115, 2016.
11. Prateek Mishra, Anish Muttreja, and Niraj K Jha. FinFET Circuit Design. In *Nanoelectronic Circuit Design*, pages 23–54. Springer, 2011.
12. Anish Muttreja, Niket Agarwal, and Niraj K Jha. CMOS Logic Design with Independent-gate FinFETs. In *Computer Design, 2007. ICCD 2007. 25th International Conference on*, pages 560–567. IEEE, 2007.
13. Farhana Sheikh and Vidya Varadarajan. The Impact of Device-width Quantization on Digital Circuit Design Using FinFET Structures. *Proc. EE241 Spring*, 1, 2004.
14. Jie Gu, John Keane, Sachin Sapatnekar, and Chris Kim. Width Quantization Aware FinFET Circuit Design. In *Custom Integrated Circuits Conference, 2006. CICC'06. IEEE*, pages 337–340. IEEE, 2006.
15. Wen-Kuan Yeh, Wenqi Zhang, Yi Lin Yang, An-Ni Dai, Kehuey Wu, Tung-Huan Chou, Cheng-Li Lin, Kwang-Jow Gan, Chia-Hung Shih, and Po-Ying Chen. The Observation of Width Quantization Impact on Device Performance and Reliability for High-k/Metal Tri-Gate FinFET. *IEEE Transactions on Device and Materials Reliability*, 16(4):610–616, 2016.
16. Brian Swahn and Soha Hassoun. Gate Sizing: FinFETs vs 32nm Bulk MOSFETs. In *Design Automation Conference, 2006 43rd ACM/IEEE*, pages 528–531. IEEE, 2006.
17. Tsung-Lin Lee, Chih Chieh Yeh, Chang-Yun Chang, and Feng Yuan. FinFETs With Different Fin Heights, August 23 2016. US Patent 9,425,102.
18. Chi-Hung Lin, Chia-Shiang Chen, Yu-He Chang, Yu-Ting Zhang, Shang-Rong Fang, Santanu Santra, and Rung-Bin Lin. Design Space Exploration of FinFETs with Double Fin Heights for Standard Cell Library. In *VLSI (ISVLSI), 2016 IEEE Computer Society Annual Symposium on*, pages 673–678. IEEE, 2016.
19. Re-Engineering The FinFET:. http://semiengineering.com/re-engineering-the-finfet/.
20. Huajie Zhou, Yi Song, Qiuxia Xu, Yongliang Li, and Huaxiang Yin. Fabrication of Bulk-Si FinFET Using CMOS Compatible Process. *Microelectronic Engineering*, 94:26–28, 2012.
21. M-S Kim, Tom Vandeweyer, Efrain Altamirano-Sanchez, Harold Dekkers, Els Van Besien, Diana Tsvetanova, Olivier Richard, S Chew, Guillaume Boccardi, and Naoto Horiguchi. Self-aligned double patterning of $1\times$ nm finfets; a new device integration through the challenging geometry. In *Ultimate Integration on Silicon (ULIS), 2013 14th International Conference on*, pages 101–104. IEEE, 2013.
22. FinFET, Multi-Patterning Aware Place, and Route Implementation:. http://go.mentor.com/4h_c2.
23. Mastering the Magic of Multi-Patterning:. http://go.mentor.com/4gue4.
24. Yongxun Liu, Kenichi Ishii, Meishoku Masahara, Toshiyuki Tsutsumi, Hidenori Takashima, Hiromi Yamauchi, and Eiichi Suzuki. Cross-sectional Channel Shape Dependence of Short-channel Effects in Fin-type Double-gate Metal Oxide Semiconductor Field-effect Transistors. *Japanese journal of applied physics*, 43(4S):2151, 2004.
25. W. P. Maszara and M. R. Lin. FinFETs - Technology and Circuit Design Challenges. In *2013 Proceedings of the ESSCIRC (ESSCIRC)*, pages 3–8, Sept 2013.
26. Brad D Gaynor and Soha Hassoun. Fin Shape Impact on FinFET Leakage with Application to Multithreshold and Ultralow-leakage FinFET design. *IEEE Transactions on Electron Devices*, 61(8):2738–2744, 2014.
27. Andy Biddle and Jason ST Chen. FinFET Technology-Understanding and Productizing a New Transistor. *A joint whitepaper from TSMC and Synopsys*, 2013.
28. M Rashed, N Jain, J Kim, M Tarabbia, I Rahim, S Ahmed, Je Kim, I Lin, S Chan, H Yoshida, et al. Innovations in Special Constructs for Standard Cell Libraries in Sub 28nm Technologies. In *Electron Devices Meeting (IEDM), 2013 IEEE International*, pages 9–7. IEEE, 2013.
29. Chi-Shuen Lee, Brian Cline, Saurabh Sinha, Greg Yeric, and H-S Philip Wong. 32-bit Processor Core at 5-nm Technology: Analysis of Transistor and Interconnect Impact on VLSI System Performance. In *Electron Devices Meeting (IEDM), 2016 IEEE International*, pages 28–3. IEEE, 2016.

30. FreePDK45 from NCSU:. https://www.eda.ncsu.edu/wiki/FreePDK45:Contents.
31. Silvestre Salas Rodriguez, Julio C Tinoco, Andrea G Martinez-Lopez, Joaquín Alvarado, and Jean-Pierre Raskin. Parasitic Gate Capacitance Model for Triple-gate FinFETs. *IEEE Transactions on Electron Devices*, 60(11):3710–3717, 2013.
32. Ning Lu, Terence B Hook, Jeffrey B Johnson, Carl Wermer, Christopher Putnam, and Richard A Wachnik. Efficient and Accurate Schematic Transistor Model of FinFET Parasitic Elements. *IEEE Electron Device Letters*, 34(9):1100–1102, 2013.
33. C Calvin Hu. Modern Semiconductor Devices for Integrated Circuits. *Part 7: MOSFETs in ICs – Scaling, Leakage, and Other Topics*, 2011.
34. Pablo Royer, Paul Zuber, Binjie Cheng, Asen Asenov, and Marisa Lopez-Vallejo. Circuit-level Modeling of FinFET Sub-threshold Slope and DIBL Mismatch Beyond 22nm. In *Simulation of Semiconductor Processes and Devices (SISPAD), 2013 International Conference on*, pages 204–207. IEEE, 2013.
35. Pranita Kerber, Qintao Zhang, Siyuranga Koswatta, and Andres Bryant. GIDL in Doped and Undoped FinFET Devices for Low-leakage Applications. *IEEE Electron Device Letters*, 34(1):6–8, 2013.
36. Seongjae Cho, Jung Hoon Lee, Shinichi O'uchi, Kazuhiko Endo, Meishoku Masahara, and Byung-Gook Park. Design of SOI FinFET on 32nm Technology Node for Low Standby Power (LSTP) Operation Considering Gate-induced Drain Leakage (GIDL). *Solid-State Electronics*, 54(10):1060–1065, 2010.
37. Thomas Chiarella, Liesbeth Witters, Abdelkarim Mercha, Christoph Kerner, Michal Rakowski, Claude Ortolland, L-Å Ragnarsson, Bertrand Parvais, Ari De Keersgieter, Stefan Kubicek, et al. Benchmarking SOI and Bulk FinFET Alternatives for PLANAR CMOS Scaling Succession. *Solid-State Electronics*, 54(9):855–860, 2010.
38. Neil HE Weste and David Money Harris. *CMOS VLSI Design: A Circuits and Systems Perspective*. Pearson Addison-Wesley, 2005.
39. Takayasu Sakurai and A Richard Newton. Alpha-Power Law MOSFET Model and its Applications to CMOS Inverter Delay and Other Formulas. *IEEE Journal of Solid-State Circuits*, 25(2):584–594, 1990.
40. James W Tschanz, James T Kao, Siva G Narendra, Raj Nair, Dimitri A Antoniadis, Anantha P Chandrakasan, and Vivek De. Adaptive Body Bias for Reducing Impacts of Die-to-die and Within-die Parameter Variations on Microprocessor Frequency and Leakage. *IEEE Journal of Solid-State Circuits*, 37(11):1396–1402, 2002.
41. Wen-Teng Chang, Shih-Wei Lin, Cheng-Ting Shih, and Wen-Kuan Yeh. Back Bias Modulation of UTBB FDSOI, Bulk FinFET, and SOI FinFET. In *Nanoelectronics Conference (INEC), 2016 IEEE International*, pages 1–2. IEEE, 2016.
42. D. Burnett, S. Parihar, H. Ramamurthy, and S. Balasubramanian. Finfet sram design challenges. In *2014 IEEE International Conference on IC Design Technology*, pages 1–4, May 2014.
43. Witek P Maszara. Finfets: Designing for new logic technology. In *Micro-and Nanoelectronics: Emerging Device Challenges and Solutions*, pages 113–136. CRC Press, 2014.
44. B. Zimmer, S. O. Toh, H. Vo, Y. Lee, O. Thomas, K. Asanovic, and B. Nikolic. Sram assist techniques for operation in a wide voltage range in 28-nm cmos. *IEEE Transactions on Circuits and Systems II: Express Briefs*, 59(12):853–857, Dec 2012.
45. T. Song, W. Rim, S. Park, Y. Kim, G. Yang, H. Kim, S. Baek, J. Jung, B. Kwon, S. Cho, H. Jung, Y. Choo, and J. Choi. A 10 nm finfet 128 mb sram with assist adjustment system for power, performance, and area optimization. *IEEE Journal of Solid-State Circuits*, 52(1):240–249, Jan 2017.
46. Woojoo Lee, Yanzhi Wang, Tiansong Cui, Shahin Nazarian, and Massoud Pedram. Dynamic Thermal Management for FinFET-based Circuits Exploiting the Temperature Effect Inversion Phenomenon. In *Proceedings of the 2014 international symposium on Low power electronics and design*, pages 105–110. ACM, 2014.
47. David Wolpert and Paul Ampadu. Temperature effects in semiconductors. In *Managing temperature effects in nanoscale adaptive systems*, pages 15–33. Springer, 2012.

48. Warin Sootkaneung, Sasithorn Chookaew, and Suppachai Howimanporn. Combined Impact of BTI and Temperature Effect Inversion on Circuit Performance. In *Asian Test Symposium (ATS), 2016 IEEE 25th*, pages 310–315. IEEE, 2016.
49. Yazhou Zu, Wei Huang, Indrani Paul, and Vijay Janapa Reddi. T_i-states: Processor Power Management in the Temperature Inversion Region. In *Microarchitecture (MICRO), 2016 49th Annual IEEE/ACM International Symposium on*, pages 1–13. IEEE, 2016.
50. Katayoun Neshatpour, Wayne Burleson, Amin Khajeh, and Houman Homayoun. Enhancing Power, Performance, and Energy Efficiency in Chip Multiprocessors Exploiting Inverse Thermal Dependence. *IEEE Transactions on Very Large Scale Integration (VLSI) Systems*, 2018.
51. Ermao Cai and Diana Marculescu. TEI-turbo: Temperature Effect Inversion-aware Turbo Boost for FinFET-based Multi-core Systems. In *Proceedings of the IEEE/ACM International Conference on Computer-Aided Design*, pages 500–507. IEEE Press, 2015.
52. Ermao Cai, Dimitrios Stamoulis, and Diana Marculescu. Exploring Aging Deceleration in FinFET-based Multi-core Systems. In *Computer-Aided Design (ICCAD), 2016 IEEF/ACM International Conference on*, pages 1–8. IEEE, 2016.
53. Emanuele Baravelli, Malgorzata Jurczak, Nicolò Speciale, Kristin De Meyer, and Abhisek Dixit. Impact of LER and Random Dopant Fluctuations on FinFET Matching Performance. *IEEE transactions on nanotechnology*, 7(3):291–298, 2008.
54. Xingsheng Wang, Andrew R Brown, Binjie Cheng, and Asen Asenov. Statistical Variability and Reliability in Nanoscale FinFETs. In *Electron Devices Meeting (IEDM), 2011 IEEE International*, pages 5–4. IEEE, 2011.
55. Shiva Taghipour and Rahebeh Niaraki Asli. Aging Comparative Analysis of High-performance FinFET and CMOS Flip-flops. *Microelectronics Reliability*, 69:52–59, 2017.
56. Hai Jiang, SangHoon Shin, Xiaoyan Liu, Xing Zhang, and Muhammad Ashraful Alam. The Impact of Self-Heating on HCI Reliability in High-Performance Digital Circuits. *IEEE Electron Device Letters*, 38(4):430–433, 2017.
57. Semiconductor Engineering Reliability Challenges In 16nm FinFET Design:. http://semiengineering.com/reliability-challenges-16nm-finfet-design/.
58. Mark LaPedus. Interconnect challenges rising: Resistance and capacitance drive need for new materials and approaches. http://semiengineering.com/interconnect-challenges-grow-3/, 2016.
59. Ning Lu and Richard A Wachnik. Modeling of Resistance in FinFET Local Interconnect. *IEEE Transactions on Circuits and Systems I: Regular Papers*, 62(8):1899–1907, 2015.
60. Spencer Tu. Putting the Pieces Together in the Materials Space: Advanced Materials Solutions for 10nm and Beyond. In *SEMICON Taiwan*, 2015.
61. Alice Wang, Anantha P Chandrakasan, and Stephen V Kosonocky. Optimal Supply and Threshold Scaling for Subthreshold CMOS Circuits. In *VLSI, 2002. Proceedings. IEEE Computer Society Annual Symposium on*, pages 7–11. IEEE, 2002.
62. Nathaniel Pinckney, Lucian Shifren, Brian Cline, Saurabh Sinha, Supreet Jeloka, Ronald G Dreslinski, Trevor Mudge, Dennis Sylvester, and David Blaauw. Near-threshold Computing in FinFET Technologies: Opportunities for Improved Voltage Scalability. In *Proceedings of the 53rd Annual Design Automation Conference*, page 76. ACM, 2016.
63. O Weber. FDSOI vs FinFET: differentiating device features for ultra low power & IoT applications. In *IC Design and Technology (ICICDT), 2017 IEEE International Conference on*, pages 1–3. IEEE, 2017.
64. Dimitrios Serpanos and Marilyn Wolf. IoT Devices. In *Internet-of-Things (IoT) Systems*, pages 17–23. Springer, 2018.
65. Gopal Singh Jamnal, Xiaodong Liu, Lu Fan, and Muthu Ramachandran. Cognitive Internet of Everything (CIoE): State of the Art and Approaches. In *Emerging Trends and Applications of the Internet of Things*, pages 277–309. IGI Global, 2017.
66. Byungseok Kang, Daecheon Kim, and Hyunseung Choo. Internet of Everything: A large-scale autonomic IoT gateway. *IEEE Transactions on Multi-Scale Computing Systems*, 2017.
67. Massimo Alioto. Enabling the Internet of Things: From Integrated Circuits to Integrated Systems. Springer, 2017.

68. Brian Bailey. Chip aging becomes design problem. *Semiconductor Engineering*, 2018.
69. James H Stathis, M Wang, RG Southwick, EY Wu, BP Linder, EG Liniger, G Bonilla, and H Kothari. Reliability challenges for the 10nm node and beyond. In *Electron Devices Meeting (IEDM), 2014 IEEE International*, pages 20–6. IEEE, 2014.
70. Christian Schlünder et al. On the influence of BTI and HCI on parameter variability. In *Reliability Physics Symposium (IRPS), 2017 IEEE International*, pages 2E–4. IEEE, 2017.
71. Ann Mutschler. Transistor Aging Intensifies At 10/7nm And Below. https://semiengineering. com/transistor-aging-intensifies-10nm/, 2017. [Online; accessed 13-July-2017].
72. Ed Sperling. Chip Aging Accelerates. *Semiconductor Engineering*, 2018.
73. Jacopo Franco, Salvatore Graziano, Ben Kaczer, Felice Crupi, L-Å Ragnarsson, Tibor Grasser, and Guido Groeseneken. Bti reliability of ultra-thin eot mosfets for sub-threshold logic. *Microelectronics Reliability*, 52(9):1932–1935, 2012.
74. Jeff Sather. Battery technologies for iot. In *Enabling the Internet of Things*, pages 409–440. Springer, 2017.
75. Scott Lerner and Baris Taskin. Workload-aware ASIC flow for lifetime improvement of multi-core IoT processors. In *Quality Electronic Design (ISQED), 2017 18th International Symposium on*, pages 379–384. IEEE, 2017.
76. Ajith Sivadasan, S Mhira, Armelle Notin, A Benhassain, V Huard, Etienne Maurin, F Cacho, L Anghel, and A Bravaix. Architecture-and workload-dependent digital failure rate. In *Reliability Physics Symposium (IRPS), 2017 IEEE International*, pages CR–8. IEEE, 2017.
77. Victor van Santen et al. Reliability in Super-and Near-Threshold Computing: A Unified Model of RTN, BTI, and PV. *TCAS-I*, 2017.
78. Jayavardhana Gubbi, Rajkumar Buyya, Slaven Marusic, and Marimuthu Palaniswami. Internet of Things (IoT): A vision, architectural elements, and future directions. *Future generation computer systems*, 29(7):1645–1660, 2013.
79. Davide Rossi, Francesco Conti, Andrea Marongiu, Antonio Pullini, Igor Loi, Michael Gautschi, Giuseppe Tagliavini, Alessandro Capotondi, Philippe Flatresse, and Luca Benini. Pulp: A parallel ultra low power platform for next generation iot applications. In *Hot Chips 27 Symposium (HCS), 2015 IEEE*, pages 1–39. IEEE, 2015.
80. T Grasser, M Waltl, G Rzepa, W Goes, Y Wimmer, A-M El-Sayed, AL Shluger, H Reisinger, and B Kaczer. The "permanent" component of nbti revisited: Saturation, degradation-reversal, and annealing. In *Reliability Physics Symposium (IRPS), 2016 IEEE International*, pages 5A–2. IEEE, 2016.
81. Minki Cho, Stephen T Kim, Carlos Tokunaga, Charles Augustine, Jaydeep P Kulkarni, Krishnan Ravichandran, James W Tschanz, Muhammad M Khellah, and Vivek De. Postsilicon voltage guard-band reduction in a 22 nm graphics execution core using adaptive voltage scaling and dynamic power gating. *IEEE Journal of Solid-State Circuits*, 52(1):50–63, 2017.

Part V
Summary and Closing Remarks

Chapter 7
Future Directions in Self-healing

7.1 Summary of the Book

The primary goal of this book is to present an effective active recovery solution to completely reverse the effect of both BTI and EM wearout. The book provides experimental demonstrations, on-chip implementations, and design methodologies for the accelerated self-healing techniques, which can act as a new dimension for mitigating wearout issues on top of any existing solutions (as shown in Fig. 7.1). Since accelerated self-healing is orthogonal to other wearout mitigation techniques and can be implemented at a relatively low cost, it can potentially be integrated with other techniques to further improve reliability and balance the trade-offs with performance and power. This book aims to build the infrastructure for exploring these methods by providing experimental evidence, circuit IP blocks, and trade-off analyses. The book also considers wearout effects in advanced FinFET nodes in emerging applications such as in the IoT domain. In summary, the book contributes to the practice of reliable system design with the following achievements:

- Performed intensive hardware measurements on FPGAs and on-chip metal lines to understand recovery behaviors for BTI and EM wearout. The measurement results and conclusions can be used to develop accurate recovery models for both wearout effects (Chaps. 2 and 3).
- Developed the BTI gate level analytic models that can be used to predict recovery rates, and can be used together with higher level models to project system resilience (Chap. 2).
- Demonstrated that both BTI and EM recovery can be performed in an active way, and the irreversible components can be completely cancelled. These properties

© Springer Nature Switzerland AG 2020
X. Guo, Mircea R. Stan, *Circadian Rhythms for Future Resilient Electronic Systems*, https://doi.org/10.1007/978-3-030-20051-0_7

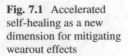

Fig. 7.1 Accelerated self-healing as a new dimension for mitigating wearout effects

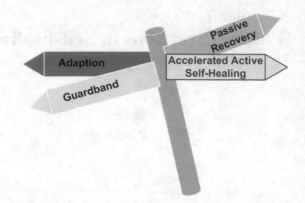

can translate into huge benefits such as reducing the need for design margins, reducing the tracking power overhead and lower area. Recovery can potentially lead to a wearout-aware and accelerated self-healing system (Chaps. 2 and 3).

- Implemented and instrumented accelerated self-healing on chip by designing a full set of circuit blocks that are able to activate and accelerate both BTI and EM recovery. A compact circuit scheme for assisting both BTI and EM recovery and supporting multiple recovery modes was also designed (Chap. 4).
- Several novel types of wearout sensors were designed. Two of them are for tracking BTI wearout, in which one is designed for separating the effect of PBTI and NBTI, and the other is designed to monitor BTI wearout and recovery. The designs can be incorporated into standard synbook flow (Chap. 4).
- Provided a series of potential cross-layer implementations especially at the architecture and system level. These solutions are able to utilize some of the intrinsic sleep behaviors for achieving self-healing. Recovery-driven design methodologies were explored to enable implementations that lead to a full accelerated self-healing system. Potential overheads are also considered (Chap. 5).
- Performed a comprehensive study across multiple technology nodes and identified the design challenges for FinFET digital circuits. Key findings included thermal effect inversion, short-channel effects, logic efforts, variations and reliability, body effect, and more. This study can function as an educational material and contribute to the growing knowledge base and design experiences for designers who are adapting to the new technologies (Chap. 6).
- Performed a complete study on impact of circuit wearout in IoT applications. The results indicated that wearout issues are very critical in several IoT domains as it is an important factor for determining the overall IoT system lifetime. This study can potentially guide IoT circuit and system designers on making high-level decisions on whether wearout effect need to be addressed for the target applications (Chap. 6).

7.2 Future Directions

For the foreseeable future CMOS technology is still likely to be the most robust and cost-effective way to implement integrated circuits. Technology scaling, although at a slower rate, still continues to be an engine of progress for the semiconductor industry. Wearout issues will continue to be a problem, and the increasing demands for robust operations within an extended lifetime by many emerging applications such as autonomous driving, medical health, and robots will increase the need for wearout-mitigation solutions as those presented in this book. Moving forward, there are many directions that can benefit or be inspired from the research results of this book to ensure a reliable system. In this section, we list a few such candidates.

7.2.1 Accelerated and Active Self-healing in Emerging Technologies

This book mainly looked into recovery behaviors in silicon-based CMOS circuits and interconnects. There has been increasing interest in other, non-CMOS, emerging technologies, specifically for non-volatile memories (NVM) [1]. Examples of such technologies are spin-transfer torque magnetic RAM (STTRAM), resistive RAM (RRAM), phase-change RAM (PCRAM), 3D-XPoint, Nantero NRAM, etc. that promise high performance, low-power consumption, and reasonable endurance. These devices are likely to exhibit their own versions of wearout, but it is likely that some fundamental similarities with CMOS wearout mechanisms will hold. For example, for STT-MRAM, wearout is due to the repeated tunneling through the magnetic tunnel junction (MTJ) with the large current during write. This is somewhat similar to BTI where voltage stress leads to the charge trapping. The notion of "accelerated self-healing" is likely to be applicable to emerging technologies and this is certainly a direction that can be further explored. As it is still debatable which level in the memory hierarchy stack is the best match for these technologies, there is a good opportunity to consider wearout and recovery in the planning stages and make optimal decisions to ensure the reliability of the overall systems.

7.2.2 Exploring Other Sources for Recovery Acceleration and Activation

The fundamentals of accelerated and active recovery are that the external stimuli (such as high temperature and reverse bias demonstrated in this book) inject energy into the device, affect the energy levels of the atoms or charge carriers, and enable a reversal of the wearout changes due to stress. Ultraviolet (UV) light is such a source of energy that has been used in the past for erasing erasable programmable read-only memories (EPROM) [2]. The charge mechanism here is the photoelectric effect where the photons directly hit and scatter the electrons from the floating gate. Therefore UV recovery has a high likelihood of removing most of the trapped charges and completely rejuvenating the flash memory or CMOS circuits. Moreover, typical EPROM erasure takes only on the order of minutes to complete—therefore we expect the time to remove the recoverable trapped charges to be of the same order of magnitude. UV recovery can be compared to thermal recovery except that the energy levels are higher, thus the recovery is likely to be faster.

7.2.3 Integrating Wearout and Recovery in EDA Design Flows

Most of the current EDA tools and flows focus on optimizing timing, power, and area. There are few tools that focus on optimizing the design for robustness. In most of the designs, wearout is addressed mainly by guardbanding, which can lead to over-engineering the system. As demonstrated in this book, recovery is very effective to exploit some of these behaviors and design for resilience. The very first step is to develop the device-level models that capture all the stress and recovery behaviors described here; these models can then be used for circuit simulation, or can be instrumented in cell libraries so that the wearout and recovery information is carried throughout the whole flow. This book can also serve as an experimental evidence for validating the models or design flows.

7.2.4 Dynamic Wearout Management by Self-learning

As we expect that wearout and thermal sensors will be distributed across the whole chip, an interesting direction is to develop learning algorithms that are able to predict the system circadian rhythms based on the history data collected from the sensors. Instead of instrumenting proactive recovery with fixed periods, these learning algorithms can guide the scheduler during run time. For example, Sarma et al. [3] made some attempts at the system level. Another direction is to explore how to adjust the dynamic thermal management policies [4] to enable thermal recovery without hurting other metrics. In the past, most of these policies focused on reducing temperature to alleviate wearout, there hasn't been efforts on recovery aspects yet. So there are good opportunities to combine dynamic thermal and wearout management techniques for enabling recovery in a smarter way.

7.2.5 Teaching Wearout and Recovery as Part of the VLSI Design Courses

There are only very few examples of covering CMOS wearout in traditional VLSI textbooks. Recovery of these wearout effects is covered even less. As reliability becomes as important as the power, performance, and area metrics, it is important to teach some aspects of wearout (e.g., trade-offs, design techniques, and recovery aspects) for students to understand the basic issues. Since this book has demonstrated several new experimental results based on modern computing platforms such as FPGAs and processors, and has also looked into wearout issues in emerging applications like IoT with the advanced nodes such as FinFET, it can serve as primary educational materials for transferring the knowledge to the classroom.

References

1. Geoffrey W Burr, Robert M Shelby, Abu Sebastian, Sangbum Kim, Seyoung Kim, Severin Sidler, Kumar Virwani, Masatoshi Ishii, Pritish Narayanan, Alessandro Fumarola, et al. Neuromorphic computing using non-volatile memory. *Advances in Physics: X*, 2(1):89–124, 2017.
2. Bert L Allen and A Rahim Forouhi. Eprom with ultraviolet radiation transparent silicon nitride passivation layer, May 12 1987. US Patent 4,665,426.
3. S Sarma, N Dutt, N Venkatasubramanian, A Nicolau, and P Gupta. Cyberphysical system-on-chip (cpsoc): Sensor actuator rich self-aware computational platform. *University of California Irvine, Tech. Rep. CECS TR-13-06*, 2013.
4. David Brooks and Margaret Martonosi. Dynamic thermal management for high-performance microprocessors. In *High-Performance Computer Architecture, 2001. HPCA. The Seventh International Symposium on*, pages 171–182. IEEE, 2001.

Appendix A
An Example Flow for Automatically Placing Metastable-Element-Based BTI Sensors

In Chap. 4 (Sect. 4.4.3), we introduced a metastable-element-based BTI sensor for tracking both wearout and recovery. The sensor is small and digital based, thus it can be embedded as an IP to be included in a standard synthesis flow. In this appendix, we will demonstrate the flow by instrumenting these sensors in a Johnson counter. This flow assumes the new scan cell which includes the sensors has been created by following the flow described in Sect. 4.4.3.7. Also, this flow is exercised with a Synopsys design environment, but similar methodology can be applied in Cadence environment as well.

A.1 New Top-Down Design Flow with BTI Sensor Insertion

As shown in Fig. A.1, the added steps for inserting BTI sensor IPs are during logic synthesis step. After the basic synthesis with Design Compiler (DC), the DFT flow runs to replace the flip-flops with scan cell. Following this, we need to update the netlist and replace this scan cell with the new scan cell which includes the sensor IP. During the physical design, the new scan cell IP needs to be added to the reference library so that the PnR tool is able to instantiate the design. The following section will demonstrate the flow step by step.

© Springer Nature Switzerland AG 2020
X. Guo, Mircea R. Stan, *Circadian Rhythms for Future Resilient Electronic Systems*, https://doi.org/10.1007/978-3-030-20051-0

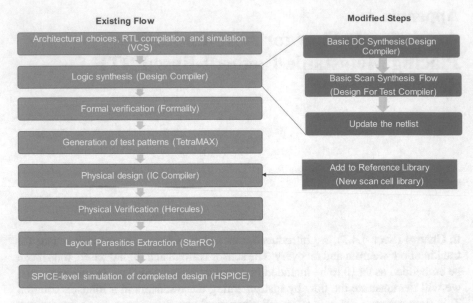

Fig. A.1 Updated top-down design flow with BTI sensor insertion

A.2   Demonstration in a Counter Design

As a proof-of-concept demonstration, we pick an 8-bit Johnson counter as our target design for sensor insertion. The detailed steps are given in Fig. A.2. After the regular DFT step, the placement strategy needs to be decided, it includes where to place the sensors, how many sensors are to be placed, and what control signals need to be added. The updated netlist with the new scan cells is shown in the bottom half of the figure, where in this case, three out of eight scan cells are replaced, the sensor inputs and control signals are also added in the signal list and are defined as inputs. After modifying the netlist from DFT, we continue P&R in IC Compiler (ICC) and the final layout with the sensor embedded is shown in Fig. A.3.

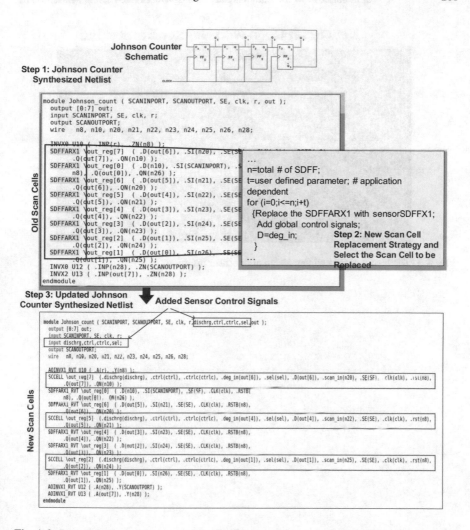

Fig. A.2 Demonstration of the sensor insertion flow in a Johnson counter design

New Scan Cells with Sensor Embedded

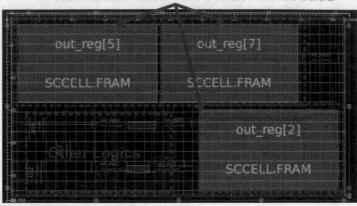

Fig. A.3 The layout of the counter design after sensor insertion

Index

© Springer Nature Switzerland AG 2020
X. Guo, Mircea R. Stan, *Circadian Rhythms for Future Resilient Electronic Systems*, https://doi.org/10.1007/978-3-030-20051-0